Asheville-Buncombe
Technical Community College
Learning Resources Center
340 Victoria Road
Asheville, NC 28801

DISCARDED

JUN 2 6 2025

Decoration and Assembly of Plastic Parts

Edward A. Muccio

The Materials Information Society

Copyright © 1999
by
ASM International®
All rights reserved

No part of this book may be reproduced, stored in a retrieval system, or transmitted, in any form or by any means, electronic, mechanical, photocopying, recording, or otherwise, without the written permission of the copyright owner.

First printing, October 1999

Great care is taken in the compilation and production of this Volume, but it should be made clear that NO WARRANTIES, EXPRESS OR IMPLIED, INCLUDING, WITHOUT LIMITATION, WARRANTIES OF MERCHANTABILITY OR FITNESS FOR A PARTICULAR PURPOSE, ARE GIVEN IN CONNECTION WITH THIS PUBLICATION. Although this information is believed to be accurate by ASM, ASM cannot guarantee that favorable results will be obtained from the use of this publication alone. This publication is intended for use by persons having technical skill, at their sole discretion and risk. Since the conditions of product or material use are outside of ASM's control, ASM assumes no liability or obligation in connection with any use of this information. No claim of any kind, whether as to products or information in this publication, and whether or not based on negligence, shall be greater in amount than the purchase price of this product or publication in respect of which damages are claimed. THE REMEDY HEREBY PROVIDED SHALL BE THE EXCLUSIVE AND SOLE REMEDY OF BUYER, AND IN NO EVENT SHALL EITHER PARTY BE LIABLE FOR SPECIAL, INDIRECT OR CONSEQUENTIAL DAMAGES WHETHER OR NOT CAUSED BY OR RESULTING FROM THE NEGLIGENCE OF SUCH PARTY. As with any material, evaluation of the material under end-use conditions prior to specification is essential. Therefore, specific testing under actual conditions is recommended.

Nothing contained in this book shall be construed as a grant of any right of manufacture, sale, use, or reproduction, in connection with any method, process, apparatus, product, composition, or system, whether or not covered by letters patent, copyright, or trademark, and nothing contained in this book shall be construed as a defense against any alleged infringement of letters patent, copyright, or trademark, or as a defense against liability for such infringement.

Comments, criticisms, and suggestions are invited, and should be forwarded to ASM International.

ASM International staff who worked on this project included Veronica Flint, Manager of Book Acquisitions, Bonnie Sanders, Manager of Copy Editing, Grace Davidson, Manager of Book Production, Alexandra Hoskins, Copy Editor, and Alexandru Popaz-Pauna, Production Coordinator.

Library of Congress Cataloging-in-Publication Data

Edward A. Muccio
Decoration and assembly of plastic parts
Includes bibliographical references and index.
1. Plastics—Finishing. 2. Plastics—Bonding. I. Title.
TP1170.M83   1999   668.4'1—dc21   99-37114

ISBN: 0-87170-634-2
SAN: 204-7586
ASM International®
Materials Park, OH 44073-0002

Printed in the United States of America

Dedication

To my son Peter and my daughter Janna.
If a father's legacy is measured through the deeds of his children...then I, indeed, have all the reason in the world to be proud.

ASM International Technical Books Committee (1998-1999)

Sunniva R. Collins (Chair)
Swagelok/Nupro Company
Eugen Abramovici
Bombardier Aerospace (Canadair)
A.S. Brar
Seagate Technology
Ngai Mun Chow
Det Norske Veritas Pte Ltd.
Seetharama C. Deevi
Philip Morris, USA
Bradley J. Diak
Queen's University
Richard P. Gangloff
University of Virginia
Dov B. Goldman
Precision World Products
James F.R. Grochmal
Metallurgical Perspectives
Nguyen P. Hung
Nanyang Technological University

Serope Kalpakjian
Illinois Institute of Technology
Gordon Lippa
North Star Casteel
Jacques Masounave
Université du Québec
Charles A. Parker
AlliedSignal Aircraft Landing Systems
K. Bhanu Sankara Rao
Indira Gandhi Centre for Atomic Research
Peter F. Timmins
Risk Based Inspection, Inc.
George F. Vander Voort
Buehler, Ltd.
A.K. Vasudevan
Office of Naval Research

Contents

Preface ix
Chapter 1: Introduction 1
   What is the Significance of Decoration and
      Assembly Processes to the Plastics Industry? 1
Chapter 2: Materials and Properties 9
   The Nature of Plastic Materials 9
   Why Should the Plastics Processor That Decorates and
      Assembles Products Understand Plastics? 10
   Feedstock Materials 11
   Materials Classification 12
      Thermosets 12
      Thermoplastics 14
   Engineered Thermoplastic Materials 17
      Modifying the Polymer 17
      Plastic Compounds 18
      Reinforcements 18
      Fillers 20
      Colorants 20
      Flame Retardants 21
      Stabilizers 21
      Plasticizers 22
      Antistatic Agents 22
      Foaming Agents 24
      Regrind 24
      Biocides 25
   The Surface of Plastics 25
Chapter 3: Adhesives 29
   Glue 29
   Adhesive 29
   Gluing 29
   Adhesion 30

Bonding Mechanisms 31
Types of Adhesives 33
Function of Adhesives 33
Adhesion Selection Process 33
    Adhesive Material Selection Matrix.................................... 34
Classification of Adhesives 35
    Function Type 35
    Chemical Families 37
Thermoset or Thermoplastic 37
Physical Form... 40
    One-Component System 40
    Two-Component or Multicomponent System 40
    Film.. 40
Assessing Adhesive Properties 40
    Adhesive... 41
    Loading Environment... 41
    Bond Strengths 42

Chapter 4: Welding Assembly of Plastics 45
Spin Welding 45
Fusion Bonding 47
Vibration Welding.. 50
Ultrasonic Welding.. 52
    Plastic Materials and Ultrasonics 55
    Ultrasonic Welding Characteristics 55
    Ultrasonic Staking 58
    Joint Design 62
    Ultrasonic Installation of Inserts
       in Thermoplastic Components 69
Ultrasonic Assembly System 73
    Converter 73
    Booster 74
    Horn 76
Dielectric Sealing 78
Induction Inserting 79
    Inserting Metal into Plastic 81
    Metal to Plastic Bonding... 84
    Simultaneous Insertion of Three Steel Inserts 91
    Inserting Metal Parts Accurately 92
    Applications 92
Induction Bonding... 99
Heat Staking 100
    Modern Heat Staking: A Comparison of Three Methods......... 100
    Advantages and Disadvantages..................................... 104

    Post Features 105
    Stake Head 105
    Side Swaging 109
  Hot Gas Welding 110

Chapter 5: Hot Stamping 113
  Advantages of Hot Stamping 113
  Hot Stamping Process 116
  Vertical Stamping Technology 120
  Hot Stamping Foils 129
    Metallic Foil Construction 130
    Pigment Foil Construction 131
    Peripheral Marking Technology 132
    Roll-on Decorating Technology 136

Chapter 6: Pad Printing 141
  Advantages of Pad Transfer Printing 141
  Pad Printing Process 143
  Vertical Printing Technology 145
  Engraved Plate Technology 148
  Doctor Blade Technology 150
  Transfer Pad Technology 151
  Ink Technology 153
  Part-Holding Fixture Technology 155
  New Viscosity Control Technologies 157

Chapter 7: Metallization of Plastics 161
  Vacuum Metallization 161
    Vacuum Metallization Process 165
    Tests 168
    Costs 170
  Electroless and Electrolytic Plating of Plastics 170
    Selecting a Plastic Material 171
    Preplate 172
    Electrolytic Plating 175
    Racking Plastic Parts for Plating 177
    Measuring Quality 178
    Design Issues and Concerns 179

Chapter 8: Painting, Coating, and Printing 181
  Internal Coloring 181
    Precolored Plastic 182
    In-House Coloring 182
  Painting 183
    Understanding the Properties of the Substrate 183
    Types of Paint 183

- Paint System Components ... 187
- Conformal Coatings ... 188
- Application Techniques ... 189
  - Spray Painting Basics ... 190
  - Electrostatic ... 190
  - Dipping ... 192
- Printing ... 193
  - Laser Printing/Etching ... 193
  - Spray and Wipe ... 195
  - TAFA ... 195
  - Screen Printing ... 196
  - Ink Jet Printing ... 197

Chapter 9: Surface Preparation ... 199
- Corona Discharge ... 199
  - Basics of Surface Modification with Electrical Discharges ... 200
  - Electrical Discharge Treatment Equipment ... 202
  - Other Treatment Issues ... 204
- Flame Treatment ... 204
  - Theory ... 209
  - Gas/Air Mixture Control ... 210
  - Flame Geometry ... 211
  - Electronic Control ... 214
- Plasma ... 215
  - Plasma Surface Treatment ... 215
  - Effectiveness of Plasma Processes ... 219
  - Engineering Plastics ... 220
  - Commodity Resins ... 221
  - Composites ... 222
  - Conclusions ... 222
- Chemical ... 223

Chapter 10: Deflashing and Cleaning of Plastic Parts ... 225
- Degating ... 225
- Flash Removal ... 227
  - Cutting and Trimming ... 228
  - Tumbling ... 229
  - Media Blasting ... 230
  - Cryogenics ... 231
- Cleaning ... 231
  - Soaps and Detergents ... 232
  - Degreasers ... 232
- Activated Gas Plasma Cleaning ... 234
  - Environmental Effects ... 235

Index ... 237

Preface

In the process of manufacturing plastic products, a significant amount of time, energy, and money is spent developing the primary processing operation that will be utilized to produce the plastic product. All this effort is, of course, critical to creating a quality plastic product. Decorating and assembly processes have often been categorized as secondary or finishing operations, and as such, they often received secondary consideration as to issues of design, equipment requirements, cost, and quality.

It is the purpose of this book to highlight and document the decorating and assembly processes in a clear and direct manner, thus affording all processors and designers of plastic parts an opportunity to correctly specify and utilize these processes.

One of the truths associated with the manufacturing of plastic parts is that the majority of plastic products require some form of decorating and/or assembly operations. It may be that the product has a snap-fit assembly, is electroplated, is assembled using adhesives, is chrome plated, or simply has the need for a manufacturer's date code. Regardless of the level of technology employed in the decorating and assembly operation, it can be confidently stated that the cost of quality defects as a result of an inadequate secondary operation will be significantly greater than that of a quality problem in the primary side of the manufacturing process. The reasoning is simple, decorating and assembly operations include the entire cost of the primary manufacturing process.

This book is directed toward individuals with little or no formal plastics education. The book should be of significant use toward such activities as process/project technicians and engineers, manufacturing operators and supervisors, purchasing and quality assurance personnel, and design engineers.

Acknowledgments

I would like to acknowledge the time, effort, support, and advice of the following individuals who reviewed various chapters of this book: Larry Schult, Professor, Plastics Engineering Technology Programs, Ferris State University, Big Rapids, Michigan; Dave Block, Manager, Dott Industries, Deckerville, Michigan; Tim Hondrop, Lacks Industries—Airlane Plant, Grand Rapids, Michigan; Mike Kemen, Plastics Resource Center Manager, Attwood Company, Lowell, Michigan. The time they spent and the experience they shared is sincerely appreciated.

Additionally, I would like to thank the many individuals and companies that provided information and materials for this book.

Finally, I would like to thank all my students in the Plastics Product Design and Decorating and Assembly classes at Ferris State University for their support and enthusiasm.

1

Introduction

What is the Significance of Decoration and Assembly Processes to the Plastics Industry?

Virtually all plastic products have some form of decoration and assembly operation involved. Whether it is the simple marking of a date code on a product (Fig. 1.1), a very sophisticated electroplating of a grill for a new pickup truck design (Fig. 1.2), the simple snapping of two plastic parts together (Fig. 1.3), or the very sophisticated ultrasonic welding of two plastic parts used in a critical medical application (Fig. 1.4), decoration and assembly is an important factor in the manufacturing of plastic parts.

Decoration and assembly processes, ironically, are not as well understood as the primary plastic processes, such as injection molding, extrusion, blow molding, or thermoforming. This lack of understanding may result in expensive scrap and a delivery problem that exceeds any scrap problems associated with the primary processes.

Consider the basic cost model for a plastic part that requires a decoration and/or assembly operation (see Fig. 1.5). Assume a company uses injection molding for an exterior part of a new sedan. The molded part is painted and assembled to a different plastic part using the ultrasonic welding process. The completed assembly has a high visibility and may affect the customer's perception of the quality of the vehicle.

The injection molded part has a standard cost equal to the total cost of the material (M), labor (L), and overhead (O) for the primary process. The material is the plastic pellets used to mold the part; the labor is any direct labor, such as molding press operator's time; and the overhead is the cost of

2 / Decoration and Assembly of Plastics

the manufacturing infrastructure, such as depreciation, utilities, and employee-related benefits. If the injection molded part has a defect and is deemed not worthy of additional operations, it will either be reworked or possibly recycled by being granulated and remolded. Although the labor and overhead cost component will be expended twice, the material cost component may be recovered. Depending on the size of the plastic part, the cost of the material, and the specific process used, the material cost can be between 25 and 70% of the standard cost to make the part.

Carrying the process forward, acceptable molded parts then have the necessary decoration and assembly processes applied. The standard cost structure for this phase of the operation uses the entire primary process standard cost (M + L + O) as part of the material component of the decoration and assembly standard cost. Additionally, paint and supplies are included in the material cost component. The labor and overhead components are related to the decoration and assembly processes. These processes are referred to as "value-added" processes.

It is when the decorated and assembled part is deemed inadequate in terms of quality that things get really expensive. Unlike the primary process, there

Fig. 1.1 Date code

are significantly fewer opportunities to recycle/reuse decorated and/or assembled parts. If repair or recovery of the decorated or assembled part is not possible, the resulting scrap cost is significant because it includes the standard cost of both the primary and secondary operations. Additionally, disposing of finished parts may be more expensive because of environmental and landfill regulations.

Fig. 1.2 Chrome-plated truck grill

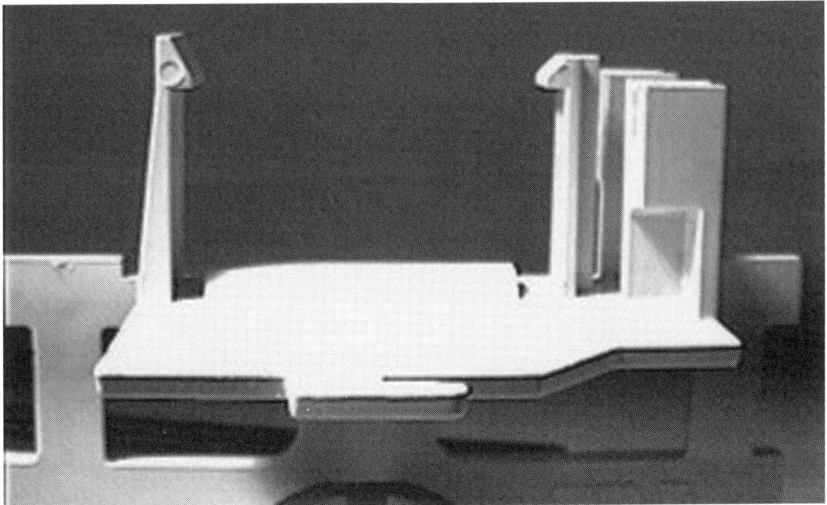

Fig. 1.3 Snap-fit part

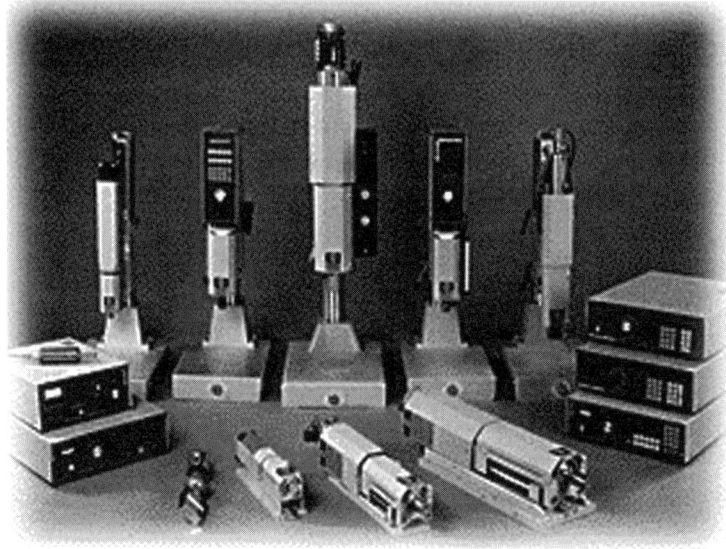

Fig. 1.4 Ultrasonic welding equipment

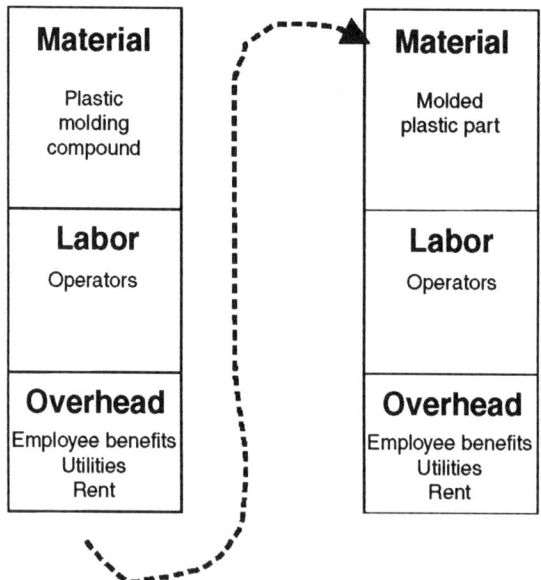

Fig. 1.5 Primary and secondary process costs

A large portion of this potential cost can be avoided if engineers, technicians, and manufacturers understand the decoration and assembly processes, including their capabilities and limitations.

Equipment used in the decoration and assembly of plastic parts is unique. Although much of the early technology was adapted from the metal, paper, and wood industries, the modern decoration and assembly systems (Fig. 1.6) are designed specifically for plastics and demand that the user understand the limits of both the equipment and the material on which the equipment is used.

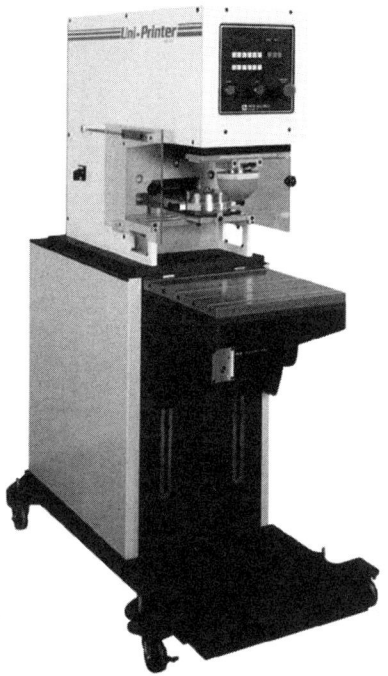

Fig. 1.6 Decorating equipment

6 / Decoration and Assembly of Plastics

Safety and Legal Factors. The plastics industry has always been concerned that new manufacturing techniques respect the health and safety of the individuals producing the products, the end user who purchases the products, and the environment where these products are used and disposed when their useful life has ended.

The Clean Air Act of 1990 caused many manufacturers of painted and coated products to invest in more efficient and environmentally safe equipment. Some manufacturers opted not to make such investments and have abandoned various painting and coating operations. Still other manufacturers have relocated processes to locations with fewer regulatory controls to avoid large capital investments in order to meet strict environmental emission standards.

On the positive side of these ever-increasing environmental regulations, paint and coating manufacturers have developed materials with low or no volatile organic compounds (VOC) that meet or exceed all the stringent requirements of the end-use applications. Advances have been made in alternative processes, such as hot stamping and thermal transfer marking, that eliminate the need for many plastic product manufacturers to use paints, coatings, and solvents altogether.

The impact and importance of decoration and assembly of plastic products can be easily comprehended by examining a common product.

Fig. 1.7 Compact disc (CD)

The compact disc (CD) is one of the most ubiquitous products of the 1990s and will continue to grow in use (Fig. 1.7). The CD is used for audio reproduction, video reproduction, and computer software storage. Even now, it is challenging to think of our lives without the CD.

When we buy and use an audio CD of our favorite music this is what we get: First of all, an injection molded jewel case, which remains closed by means of a subtle interference snap, opens employing a simple nesting hinge. Next, the CD is securely supported in the case by means of an annular snap, which fits to lock in the center of the CD. The CD itself works because of the reflectance of the vacuum metallized surface. The CD is labeled using artistic graphics that were either hot stamped or pad printed onto the CD itself. Finally, either the CD, the jewel case, or both are labeled with a holographic security label and printed with a stamped or laser-etched manufacturing code.

This one product is the result of several decoration and assembly operations that manufacturers and product designers need to understand to control the cost of production.

2

Materials and Properties

The Nature of Plastic Materials

Plastics are polymers with specific additives incorporated to make what is commonly called a plastic compound. A polymer is an organic (based on carbon) macromolecule (large in size) comprised of several thousand repeating segments (called "mers") that are linked together in a chainlike form (Fig. 2.1). The number of times a particular segment repeats is referred to as n, the degree of polymerization. As n becomes larger, the polymer molecule becomes longer and the molecular weight of the polymer increases. The

Polymers are:
- ▼ Organic
- ▼ Large Molecules
- ▼ Chain-Like
- ▼ Made From "Building Block" Molecules

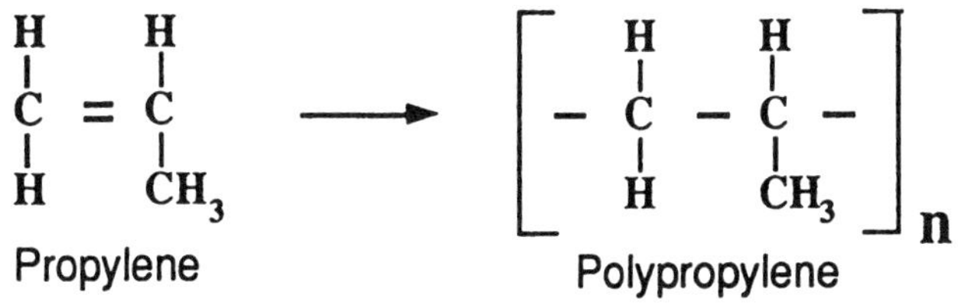

$$\underset{\text{Propylene}}{\overset{\displaystyle H \quad\; H}{\underset{\displaystyle H \quad\; CH_3}{C = C}}} \longrightarrow \left[\underset{\text{Polypropylene}}{\overset{\displaystyle H \quad\; H}{\underset{\displaystyle H \quad\; CH_3}{- C - C -}}} \right]_n$$

Fig. 2.1 Facts to remember about polymers

chainlike molecule is often referred to as the "carbon chain" (Fig. 2.2). When compared to common materials, such as water or oil, the carbon chain molecules that define a plastic are long (e.g., $n = 100$ to > 1000).

Why Should the Plastics Processor That Decorates and Assembles Products Understand Plastics?

Plastics are unique, and there are thousands of unique plastics. Understanding how plastics/polymers work helps the plastics processor make informed decisions when selecting a plastic for a particular application or establishing the best processing parameters. The unique characteristics of plastic materials require decorators and assemblers of plastic products to correctly assess the properties of a material to successfully finish the product.

To help understand the effect of the length of plastic molecules, it is helpful to think of long polymer chains as spaghetti cooking in a pot of boiling water. If the average length of each piece of pasta is 5 cm (2 in.), a cook would be hard pressed to pick up very much on a fork. On the other hand, if the average length of each spaghetti strand is 20 to 30 cm (8 to 12 in.), the cook would find that the pasta becomes easily entwined and tangled, allowing a larger amount to be gathered on the fork.

The longer the polymer molecules are, the more they become entangled. The degree of entanglement helps give mechanical properties and flow characteristics, such as melt viscosity, used when processing the plastic.

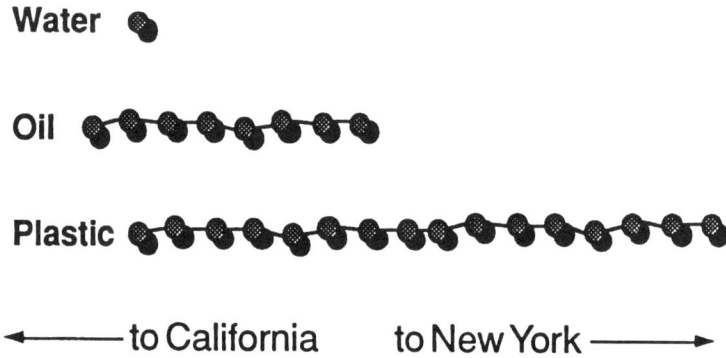

Fig. 2.2 The length of the carbon chain in a plastic compared to water and oil

Feedstock Materials

What is the source of plastics? The feedstock materials for plastics fall into three basic categories (see Fig. 2.3):

- *Petroleum:* Usually in the form of a light distillate, such as benzene
- *Natural gas:* In the form of methane
- *Agricultural materials:* In the form of wood or cotton (cellulose) or soybean by-products

Large volume plastic users (Fig. 2.4) should carefully monitor petroleum-based feedstock prices and the price of plastic materials. Political factors, environmental concerns, and natural disasters have a major impact on feedstock availability and, therefore, pricing.

The feedstock materials are chemically developed into what is called a monomer (i.e., single unit). The monomer then reacts with a catalyst, heat, and pressure to create the polymer. This process, called polymerization, is usually a batch-type process in which several thousand pounds of polymer are manufactured in large reactors.

The plastic mass has to be granulated and then pelletized to a cylindrical shape approximately 1.5 mm (0.060 in.) in diameter by 4.6 mm (0.180 in.) in length. Other plastics, such as polystyrene, can be polymerized directly

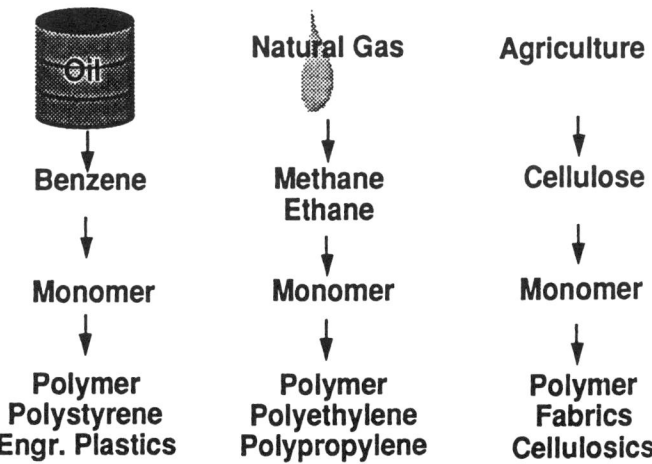

Fig. 2.3 Feedstock materials for plastics

into a spherical shape. These shapes allow the plastics processor to conveniently transport and handle the plastics in the various processing equipment (Fig. 2.5). Figure 2.5 shows the three common packaging systems in which plastic pellets are shipped: 50 lb sacks, 300 lb cardboard drums, and 1000 lb Gaylord (Gaylord Container Corp., Deerfield, IL) containers. Prior to the pelletizing process, the plastic may be compounded with various additives, which are described later in this chapter.

Materials Classification

Plastic materials are named and classified to the point where many plastic product designers become confused or overwhelmed by the variety of chemical and trade names. To make sense of the nomenclature, it is best to start with the two major plastics categories: thermosetting and thermoplastic materials (Fig. 2.6).

Thermosets

Thermosetting plastics are plastic compounds that "set" or crosslink upon heating. The crosslinking process actually is the formation of chemical bonds between the long carbon chains (Fig. 2.6). The additional chemical

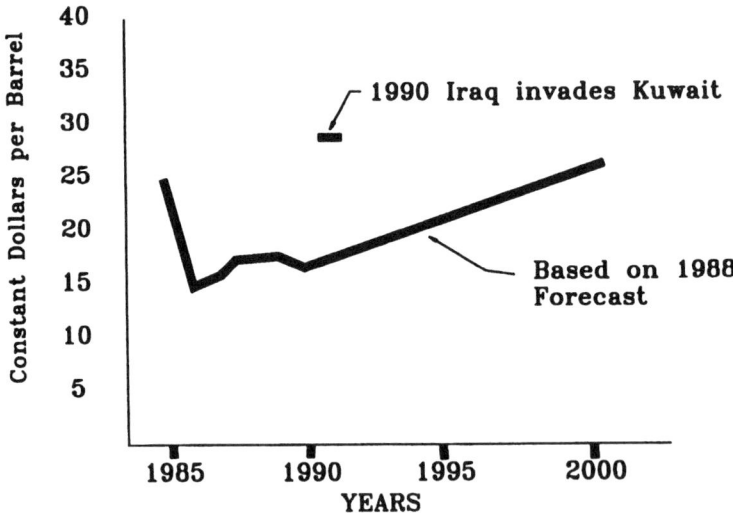

Fig. 2.4 Effect of OPEC oil pricing on the price of plastic

bonds of the crosslinked thermosetting plastic allow more thermal energy (heat) to be absorbed before the carbon chain is broken. This process allows thermosetting plastics to perform at higher temperatures and develops a material with outstanding chemical and electrical resistance.

The thermosetting process is irreversible. Once set, the plastic can not revert to its prior stage. An example often used to clarify this irreversible

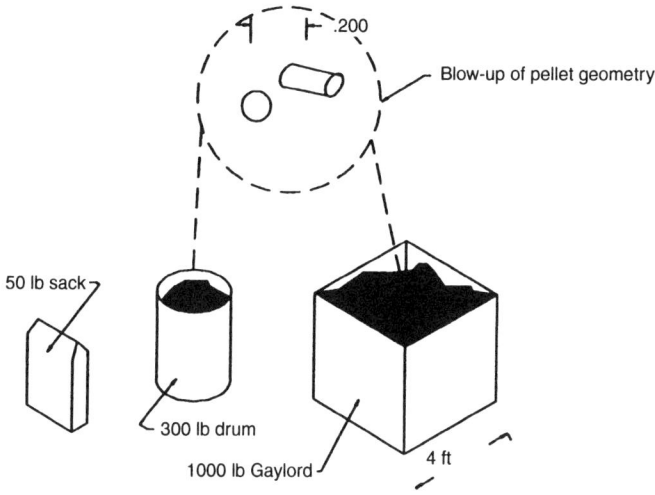

Fig. 2.5 Plastic containers and pellet shape

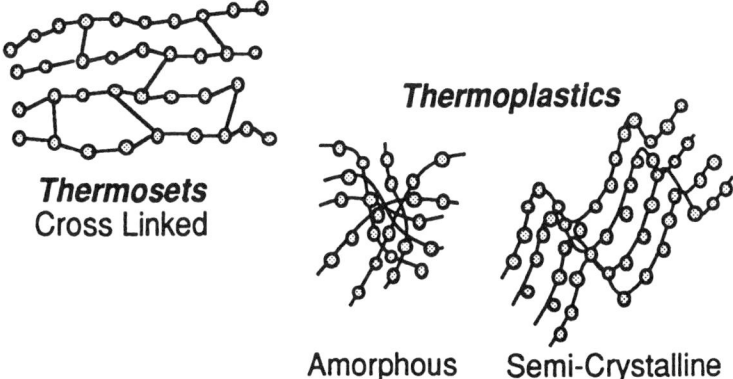

Fig. 2.6 Comparison of molecule structure in thermosets and thermoplastics

process is baking a cake. Ingredients are mixed, and the batter is baked. If the baker is not satisfied with the quality of the cake, there is no way to fix or undo the baking process. It is irreversible.

Terms often used to refer to the crosslinking process include set, cure, vulcanize, and kick over. Typical plastics that are thermoset are listed in Fig. 2.7 with respect to their usage growth.

Thermoplastics

Thermoplastics are those plastic materials that soften on heating and harden on cooling. Contrary to thermosetting, this process is reversible. An example used to describe thermoplasticity is ice, which hardens when cooled and can be remolded by heating.

Unlike thermosets, there are no chemical bonds between the long chain molecules (Fig. 2.6), but the position of thermoplastic molecules (and the intermolecular forces that hold them together) further affects their properties and classification.

Forms of Thermoplastics. Thermoplastic materials are found in a wide variety of formations, which include:

- *Amorphous:* Without logical order
- *Semicrystalline:* Molecules have order
- *Elastomer:* Materials able to stretch 2 times the original length and fully recover

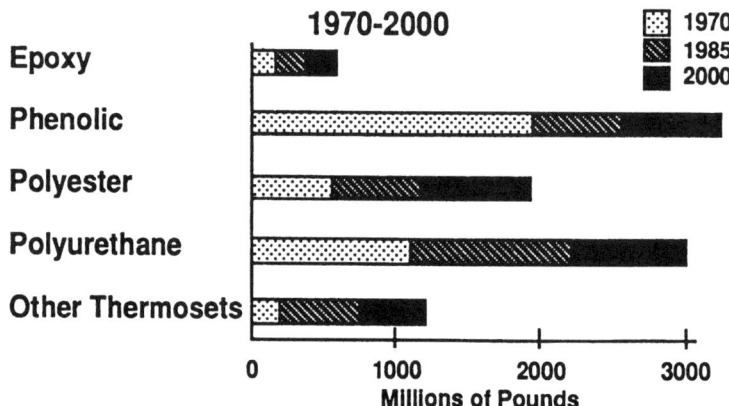

Fig. 2.7 Cumulative domestic thermoset growth. Source: "Plastics A.D. 2000," Society of the Plastics Industry

- *Plasticity:* Materials tend to remain in a new shape if distorted (the opposite of elasticity)
- *Rigid:* Materials that are stiff and maintain their shape
- *Flexible:* Materials that are easily folded or distorted
- *Ductile:* Materials that can be stretched or pressed without destroying the basic integrity of the material

Utilizing this information, polystyrene (the plastic used to make model airplanes) could be described as a rigid, amorphous thermoplastic, and low density polyethylene (the material used in plastic bags) could be described as a flexible, semicrystalline thermoplastic. It should be noted that a flexible plastic may not necessarily be an elastomer; however, virtually all elastomers are flexible.

Figure 2.8 illustrates the difference in tensile properties between plastics and metal. The stress is the force/area applied in a tensile (pulling) mode, and the strain is the resulting deformation (change in length/original length) of any materials. Stress and strain are discussed in more detail in Chapter 3.

Amorphous Thermoplastics. Random entanglement of the thermoplastic molecules (much like the long spaghetti on the end of the fork) is called amorphous structure. Amorphous means without a logical order. Amorphous thermoplastics can be clear, have uniform (isotropic) properties in all

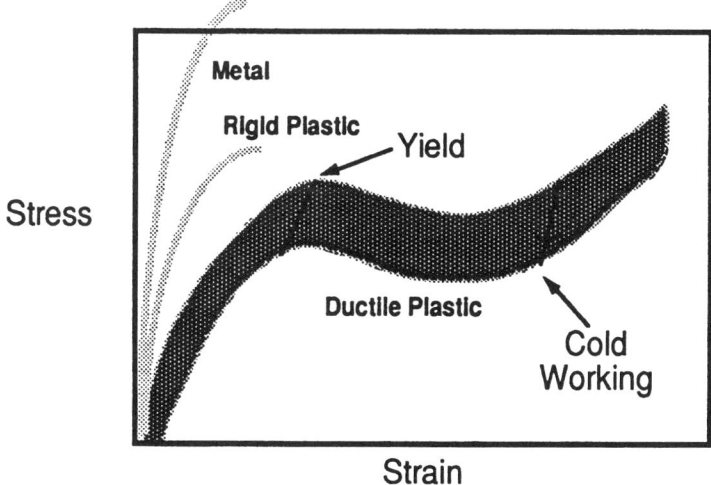

Fig. 2.8 Tensile properties of plastic versus metal

directions, and have a melting range versus a melting point. Additionally, amorphous plastic materials tend to have a lower chemical resistance than semicrystalline plastics. This chemical behavior can have a major affect on the ability of plastics to be successfully bonded, painted, or welded. Typical amorphous thermoplastics include polystyrene, acrylics, polycarbonate, and polyvinyl chloride.

Semicrystalline Thermoplastics. Molecules that have an order to them are referred to as semicrystalline (often referred to as "crystalline"). Unlike crystal structure in salt or metals, semicrystalline in thermoplastics indicates an occasional order where molecules may line up next to one another (Fig. 2.6). Its opacity, nonuniform (anisotropic) properties, and distinct or narrow melting range characterize a semicrystalline thermoplastic. Additionally, semicrystalline plastics tend to have excellent chemical resistance and a waxy surface. However, the ability of a material to be bonded, painted, or coated often is decreased. Typical semicrystalline thermoplastics include polyethylene, polypropylene, nylon, and thermoplastic polyesters. Figure 2.9 highlights a variety of thermoplastic materials by their projected growth.

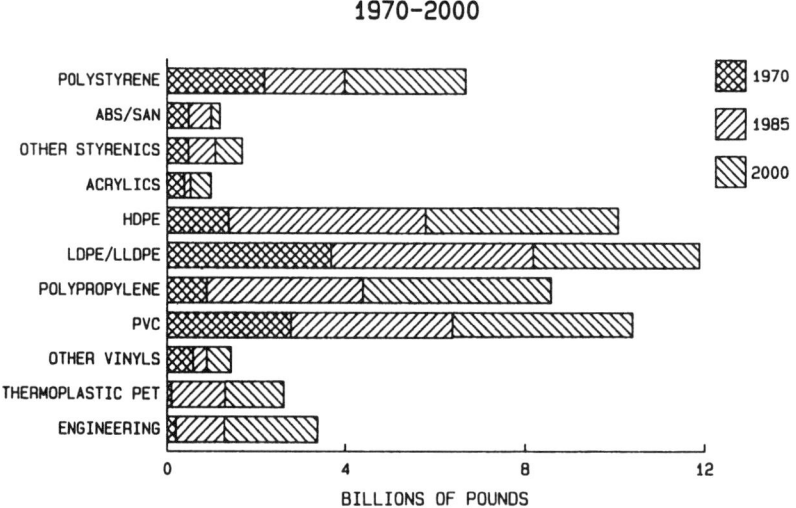

Fig. 2.9 Cumulative domestic thermoplastic growth. Source: "Plastics A.D. 2000," Society of the Plastics Industry

Engineered Thermoplastic Materials

Most of the plastics used today are not pure polymers. The requirements of a specific application and the properties of a plastic may not match; however, a polymer can be modified (or custom engineered) to improve specific properties. There are two main techniques for enhancing a plastic: modifying the polymer and creating a plastic compound.

Modifying the Polymer

If a selected polymer is deficient in one or more properties required in the end-use application, a hybrid plastic is often created by copolymerizing two or more plastics together (see Fig. 2.10), thus combining the best properties of each plastic into one material. An example of a copolymer would be ABS (polyacrylonitrile butadiene styrene), which combines three polymers. The "A" in ABS stands for acrylonitrile. Acrylic polymers are known for their clarity, colorability, and brittleness. The "B" stands for butadiene, which is a rubberlike material that acts much like a shock absorber, thus improving impact resistance. The "S" stands for styrene. Polystyrene is a clear, brittle, low cost polymer. When polymerized together, these three different polymers, called a terpolymer, exhibit a phenomenon known as synergism. Synergism occurs when the result exceeds the sum of the parts. In the case of

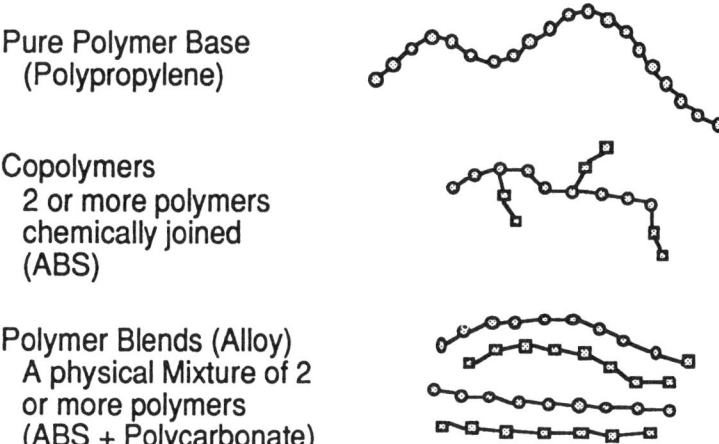

Fig. 2.10 Common techniques for custom engineering of polymers

ABS, the resulting physical properties (low cost, impact resistance, and good colorability) are found in the product, not the components.

Another technique for customizing plastics is to physically blend two or more polymers together (see Fig. 2.10). There is no chemical bonding in a blend. Often a blend of polymers is referred to as an alloy. A good example of a polymer blend is that of ABS and polycarbonate. Polycarbonate is known for high impact resistance and natural flame retardancy. Polycarbonate is also more expensive than ABS. Customers of the plastic industry needed a plastic with better impact resistance and a better flammability rating than ABS, but they could not justify the cost of polycarbonate. However, ABS and polycarbonate can be compatible (not all plastics are compatible with each other) and can be physically blended to meet customer requirements.

In addition to copolymerization and blending, polymer engineers have been able to create graft and branched polymers. Graft polymers chemically bond different materials at select points on the carbon chain to customize properties. Branched polymers have a modified carbon chain with smaller chains or branches bristling off the main chain, which provide additional strength and rigidity to the polymer.

Plastic Compounds

Even with all of the polymer engineering efforts to customize a polymer to meet specific customer needs, there is still a demand for additional property modifications that cannot be met by the polymer itself.

Most plastics used today are comprised of a polymer and additive(s). The additive modifies specific properties of the plastic to further characterize the material. These combinations of additives and polymers are referred to as plastic compounds. Plastic compounds include reinforcements, fillers, colorants, flame retardants, stabilizers, plasticizers, foaming agents, and regrind. It is unlikely that "neat" plastic (polymer only, no additives) will have all the desired properties, such as strength, density, color, thermal properties, and cost efficiency, to meet a specific application. Therefore, a plastic product designer must be familiar with what additives and modifiers are available and how they affect (positively and negatively) the plastic compound and ultimately the plastic product.

Reinforcements

Reinforcements are used to enhance specific (usually mechanical) properties, such as tensile strength or flexural modulus (rigidity). Thermal properties, such as deflection temperatures, are also augmented with reinforcements. Reinforcement materials are treated with a chemical coupling agent, which helps the reinforcement to remain attached to the plastic matrix.

Major advantages of using reinforcements include improved strength and rigidity as well as improved heat resistance. Major disadvantages include higher production costs, shorter equipment/tooling life, and reduced product surface appearance.

Glass, in the form of fibers compounded with plastic and pelletized for convenient use by plastic molders, is the most popular reinforcement material for use with plastics. Because the glass fiber is usually no longer than the plastic pellet (about 6.3 mm, or 0.250 in.), the fiber has the opportunity to be molded using the same equipment and molds as the plastic without reinforcement. The addition of glass (and most other reinforcements) significantly erodes both the molding machinery and the mold. Long term processing costs should be considered when working with heavily reinforced plastics.

Another consideration for the plastic product designer is that the introduction of a fiber into the plastic melt stream may result in fiber orientation (fibers align themselves), which creates anisotropic (uneven) physical properties (Fig. 2.11).

The two major glass fibers used in plastics today are E-glass, a low-cost fiber with good electric resistance properties, and S-glass, which is more expensive than E-glass but offers improved mechanical strength.

Carbon/graphite fibers, used predominately in advanced plastic materials, significantly improve the strength and modulus (rigidity) of a plastic.

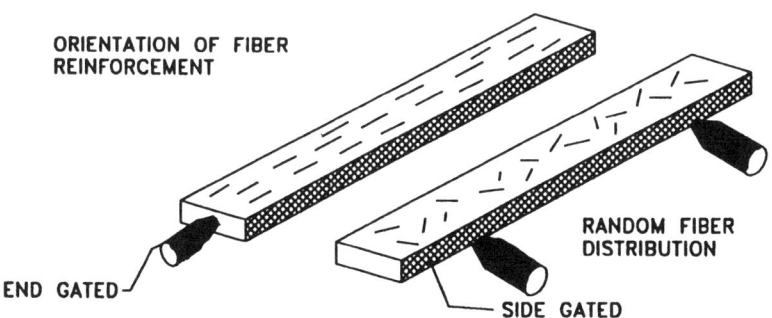

Fig. 2.11 Glass fiber reinforcement of plastic

Mica is a quartz-like mineral that is considered to be particulate in nature. A particle, unlike a fiber, is unlikely to orient during the molding process; therefore, it affords the designer isotropic (even) property behavior.

Fillers

Fillers are materials added to plastic specifically to lower cost. They are compounded and pelletized with the plastic materials. A plastic compound may contain filler(s) without the knowledge of the material processor. Typical filler materials include wood flour, kaolin (clay), and cotton and cloth. Although fillers are more commonly used with thermosetting plastics, they are available for thermoplastic materials.

Major advantages of filler materials include lower material cost, more product per pound of plastic (polymer), and improved heat resistance. Major disadvantages include lower mechanical properties than unfilled plastics (depending on the filler used) and higher process variation.

Colorants

Colorants are additives used to alter the color of plastic. The science of coloring and color matching is complex and often requires sophisticated equipment to ensure that colorants are both compatible with the plastic and are the correct type. The complexity of coloring plastic is increased because each plastic has a different color base. For example, acrylics, polystyrene, and polycarbonates are clear and, therefore, readily colored. Nylon and ABS are naturally brownish in color and are more difficult to color. Thermosetting plastic and reinforced plastic are opaque, which limits the coloring of these materials.

Colorants themselves are usually organic or inorganic dye or pigment. The amount of colorant added to plastic material is relative to the color of the base plastic (i.e., the darker the base plastic, the more colorant required). Colorant is mostly precompounded with the plastic, but colorant can be added, if justified, to the plastic by the processor. The colorant may be a powder, liquid, or pellets, and a typical colorant is added in the range of 1 to 4 wt%.

Colorants, like all additives, affect other properties. As an example, consider the coloring of plastic either white or black. White colorant is usually a form of titanium dioxide. When added to sufficiently create a good white color, the plastic part can become stiffer and suffer from a decrease in flexibility. Black colorant is usually a form of carbon black, which, when added to many plastics, increases the rigidity of the material.

Although these second order effects may be desirable, they also must be understood. A product design with a natural (uncolored) nylon plastic may perform differently than the same product manufactured with a white nylon material.

Major advantages of colorants include a wide range of available colors and coloring throughout the material. Major disadvantages include higher cost, possible negative effects on other properties, difficulty of matching colors, batch to batch color variations, and colors may change when exposed to heat and sunlight.

Flame Retardants

Plastic materials burn. Most designers want to slow the burning rate in their plastic sufficiently to meet agency and consumer requirements. Flame retardants are added to plastics to slow the rate of burning and/or create a plastic that will not be able to support a flame. Flame retardants fall into two main categories. They are either compounds that when heated generate a gas that starves a flame (i.e., removes available oxygen), or they are other more flame retardant plastics that are blended with the selected plastic.

The issues associated with plastic flammability are controversial, and careful consideration by the plastic product designer must be made to understand the value of adding a flame retardant to a plastic versus selecting a more naturally flame retardant material. Flammability ratings for plastics are somewhat subjective and confusing. They are a function of wall thickness. For example, a plastic with a 94 V-O (tested vertically) rating with a 3.2 mm (0.125 in.) wall thickness may have a different rating (more flammable) with a wall thickness of 1.6 mm (0.0625 in.).

Major advantages of flame retardants include reduced flammability and improved heat resistance. Major disadvantages include higher cost, possible processing problems, and reduced mechanical properties.

Stabilizers

Stabilizers are additives that help control or enhance specific properties. As with all additives, there may be negative aspects to their addition in a plastic system.

Thermal stabilizers are used to provide an improvement in the long-term stability of plastics when exposed to heat. Many plastics may be within a range 11 to 22 °C (20 to 40 °F) below the end use temperature. The thermal stabilizer well affords the plastic a widened temperature range required to meet an application. Thermal stabilizers are often used in heat-sensitive plastics, such as PVC.

Plasticizers

Plasticizers are chemicals added to a plastic (usually PVC) to make it flexible. Like many additives, plasticizers vary in their compatibility with the plastic compound. As a result, the plasticizer, over time, may become extracted from the plastic. A good example of this phenomenon is automobile seats. When purchasing a new car, buyers immediately become aware of the "new car" smell. This aroma is actually a combination of odors that include carpeting, adhesive, paints, solvents, and plasticizers used to keep the vinyl upholstery supple (if the seats are leather, there is still vinyl in the padded dash). Over time, repeated temperature changes, and the friction of sliding in and out of the car, the plasticizer is gradually removed from the upholstery. The "milky" mist that lingers on the inside of the windshield of a new car on a hot summer day is also a plasticizer. After enough plasticizer is extracted from the vinyl, the flexibility of the material decreases, and the seats and padded dash blister and crack.

A new market has developed to combat this problem: spray-on protectorates. These spray-on products are actually a form of plasticizer that coats the surface of vinyl preventing any further plasticizer (from within the vinyl) to be extracted.

Ultraviolet (UV) Light Stabilizers. Many thermoplastics have a tendency to fade in color or physically degrade when exposed to sunlight. The UV stabilizer provides the plastic with the needed UV resistance to be used in exterior applications. A good example of plastic that requires UV stabilization is polycarbonate. This strong thermoplastic would not be able to compete against other plastic materials in exterior applications, such as auto taillight lenses, if not for UV stabilization. Some colorants, such as carbon black, provide UV stability as a secondary benefit.

Antistatic Agents

One of the advantages of plastics is their inherent dielectric abilities. The fact that most plastics do not conduct electricity becomes a problem when static electricity needs to be dissipated. Electric charges that accumulate on the surface of a plastic remain there until neutralized. Antistatic agents fall into three major categories (Fig. 2.12): internal, external, and ion discharge antistatic systems.

Internal antistatic agents are chemicals compounded in the plastic that migrate or "bloom" to the plastic surface due to their incompatibility with the plastic. When the chemicals reach the surface of the plastic, the surface resistivity of the plastic is decreased (i.e., the surface becomes more conductive). This change in the electrical characteristics of the surface is enough to dissipate the static charge. The internal antistatic agent has a finite life, and the effect gradually disappears. Internal antistatic agents have proven helpful in the recording industry, where static charge on the surface of phonograph records and compact discs increases the attraction of dirt and dust.

External antistatic agents are applied to the surface of a plastic part after it has been manufactured. The principle of decreasing surface resistivity is the same as that of the internal antistatic agent. External antistatic agents are more effective than internal antistatic agents; however, they are more short-lived. External antistatic agents have provided a niche market with development of the anti-cling dryer sheets. Polymer fabric surface conductivity is temporarily increased, and the clothes lose their static charge.

Ion Discharge Antistatic Systems. The electronics and packaging industry had a need for short-term static charge dissipation. The use of ion sources (generated electrically or by low level radiation) has proven to be an answer. Plastic product is exposed to a slightly ionic atmosphere (such as an air stream), which has an opposite charge to the surface electrons that cause static electricity. The static charge on the plastic surface is temporally neutralized. This process is useful in the dissipation of static charges on plastic parts with complex shapes.

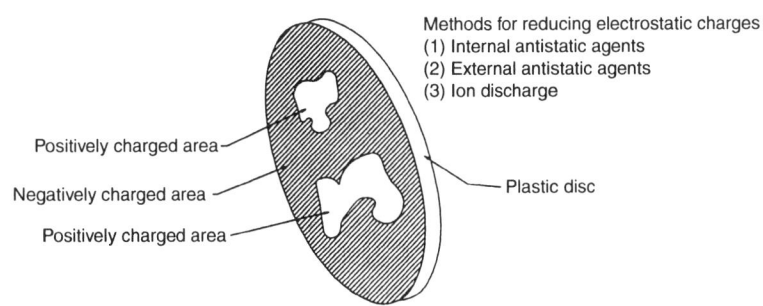

Fig. 2.12 Antistatic agents in plastic

Foaming Agents

Cellular plastic (foamed plastic) is of particular value as a thermal insulation product, servicing building and construction as well as packaging markets. All plastics can be foamed by the introduction of an additive or filler. There are three main foaming techniques:

- *Internal blowing agents:* Compounds added to the plastic decompose within a specified temperature range during molding, which results in the evolution of a gas (usually nitrogen) that when allowed to expand in the plastic melt forms a cellular structure.
- *External systems:* Usually, gases (steam or nitrogen) are physically introduced into the plastic melt to provide the cellular structure.
- *Microballoons or microspheres:* Small hollow spheres (600.25 mm, or 600.010 in., diameter) mix with the plastic (usually thermosetting plastic) that are the cell structure for a "syntactic foam."

Regrind

Thermoplastic material that has been granulated and reintroduced into the process, usually by mixing with virgin plastic, is called regrind. The introduction of regrind allows a plastic processor to get the maximum material usage; however, the regrind is not exactly the same as virgin material and may negatively affect the process. Characteristics of regrind (when compared to virgin plastic) include a longer heat exposure history, a different shape (therefore, different processing), a reduction of physical properties (Fig. 2.13), and an altered color or contamination.

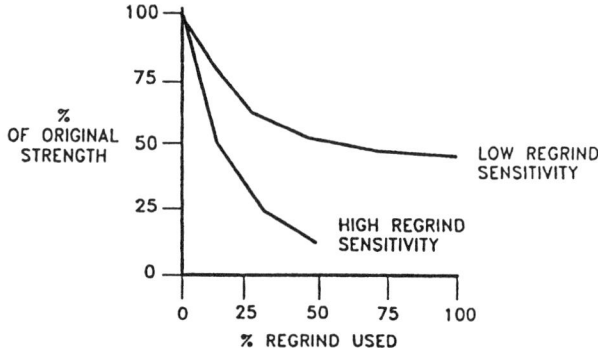

Fig. 2.13 How regrind might affect the strength of a plastic part

The plastic product designer should specify the maximum regrind allowed for a particular part directly on the print, which includes specifying "no regrind allowed" if that is what is permitted. Plastic part designers should consider the issue of regrind, not only in part design, but in the process and tooling as well. Often, plastic parts can be molded with little or no nonproduct material generated, thus not requiring regrind.

Biocides

Many plastics have a propensity to attract undesirable lifeforms, such as fungi and bacteria. Plastics have also been known as a food source for detestable creatures, such as rodents. Underground conduit and plumbing product manufacturers, in particular, have found that biocides added to plastic may provide adequate resistance and protection from these pests. A more common example of biocide use in plastics is the insecticide used in pet flea collars. This biocide migrates to the surface of the extruded vinyl collar providing the pet relief from fleas for several months.

The Surface of Plastics

Decoration and assembly of plastic parts requires an understanding of the surface of plastic materials. The varied chemical structures of plastic materials does not allow generalization of how plastics perform during secondary operations. See Chapter 7, "Surface Preparation," for more information.

As an example, a semicrystalline material, such as polyethylene or polypropylene, does not lend itself to marking with a decorating process nor allow good adhesion through adhesive systems. Amorphous plastics, such as PVC or polystyrene, readily lend themselves to decoration and assembly operations.

Whether applying a coating, such as paint, or a bonding agent, such as an adhesive, the ability for another substance to adhere to a plastic surface is a function of the ability to have intimate contact with another surface (i.e., to wet the surface). "Wetting" is defined as the displacement of air from a solid-air interface by a liquid, which creates a new liquid-solid interface. Another way to think of wetting is to visualize what occurs at the surface of a plastic part when a coating or adhesive is applied. Successful bonding requires direct contact with the plastic surface. As seen in Fig. 2.14, even the smoothest plastic surface has peaks and valleys that trap air and prevent wetting. Changing the surface chemistry of the plastic or coating, in addition to lowering the viscosity of the coating, may be necessary in order to achieve wetting.

An everyday example that describes wetting is observing water droplets on the surface of an automobile. A common indicator that it is time to wash and wax the car is when water droplets fail to bead up. After a good wash and wax job, water forms distinct beads that tend to stand high. The phenomenon of this beading is called "cissing."

Materials can also be characterized by their surface tension. Surface tension is what allows a water droplet to form in the first place, and surface tension relates to the wettability of a liquid. Water has a relatively high surface tension between 70 and 71 dynes/cm at 20 °C.

A quick experiment to illustrate this is to place a drop of water on the flat surface of a mirror that is lying face up. Next to the water droplet, put a drop of isopropyl (rubbing) alcohol (surface tension, 21 dynes/cm at 20 °C). The alcohol has a lower surface tension and spreads more than the water, which

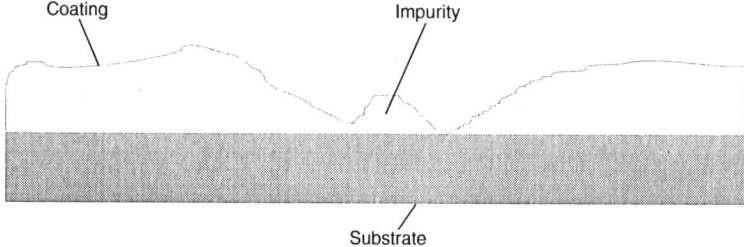

Fig. 2.14 Surface coating

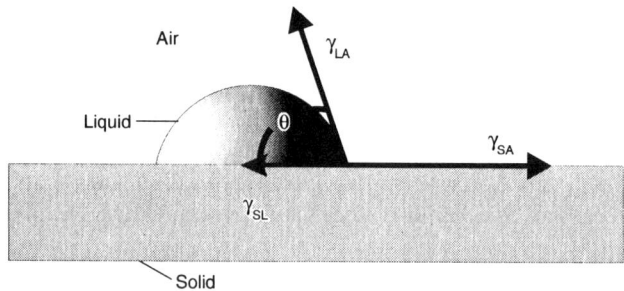

Fig. 2.15 Surface tension. γ_{SA}, interfacial free energy at the solid/air interface; γ_{SL}, interfacial free energy at the solid/liquid interface; γ_{LA}, surface tension of the liquid

is why isopropyl alcohol cleans skin more thoroughly than water. It wets the skin better. In fact, water has one of the highest surface tension values for a liquid.

Scientists describe this phenomenon in terms of the contact angle (Fig. 2.15). In order to wet the surface, the liquid must have a lower surface tension than the critical surface tension of the solid. This arrangement results in a lower contact angle, demonstrating better wetting. When coating a plastic part, beading is not desirable. The beading is caused by the high surface tension of the plastic part (or a freshly waxed car).

The challenge of an effective bond is to first be sure the coating/adhesive spreads and wets the surface of the plastic. To achieve this, the coating/adhesive must have a lower surface tension than the plastic with which it is trying to bond. If the coating/adhesive has a higher surface tension value than the plastic, one of two choices must be made:

- Lower the surface tension of the coating or adhesive with an additive or a change in temperature
- Raise the surface tension of the plastic surface with one of several possible processes

3

Adhesives

Sometimes the terminology can get in the way of understanding, which is often the case with "glue" and "adhesive."

Glue

For this discussion, glue is defined as a bonding material created from natural and formerly living things, such as horses and fish. The old adage about taking the aging horse to the glue factory has a ring of truth to it. Also, seaports often have fish glue factories nearby to process fish renderings into an amber glue used by paper manufacturers, secretaries, and school children.

Adhesive

An adhesive is a synthetic substance, usually polymeric in nature. It may be a single or multicomponent system that bonds as it polymerizes (Fig. 3.1).

Gluing

The verb *to glue* or the process of gluing is acceptable when referring to the use of glue or adhesive:

- The student glued the paper decoration together with an amber glue.
- The technician glued the panel of the B-2 Stealth Bomber with the acrylate adhesive.

Today, however, the verb *to bond* is more acceptable for referencing the use of adhesives, especially in highly technical applications.

Adhesion

The mechanism or method by which materials are bonded together can vary widely, depending on both the adhesive being used and the materials being joined. In order to clarify this explanation, some basic definitions are needed.

- *Adherend:* A body held to another body by an adhesive
- *Substrate:* A material upon the surface of which an adhesive-containing substance is spread for any purpose, such as bonding or coating. A broader term than adherend
- *Primer:* A coating applied to a surface prior to the application of an adhesive to improve the performance of a bond

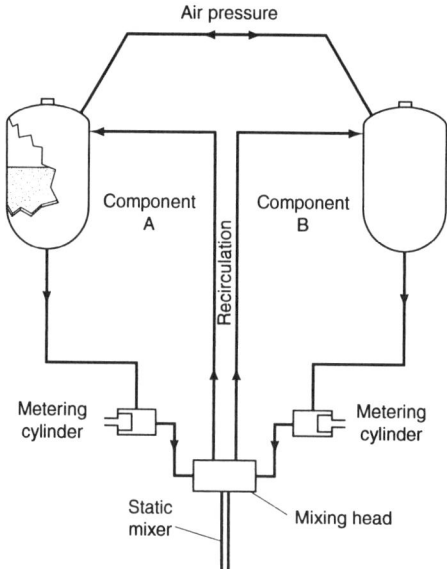

Fig. 3.1 Two component adhesive system

One unique material (Fig. 3.2) used in the plastic industry to help adhesives and paint bond to material, such as polyethylene and polypropylene, is chlorinated polyolefin (CPO). The CPO can be used either as a primer or added directly to the paint or adhesive.

Bonding Mechanisms

Bonding mechanisms can be categorized in several ways. When a carpenter hammers a nail to hold pieces of wood together, it is not considered an adhesive bond. It is, however, considered a mechanical fastening system. On a chemical basis, there is a somewhat differently defined mechanical fastening that takes place.

Mechanical and Physical Proximity. A toolmaker uses a tool called "Joe Blocks." These are highly polished and dimensionally precise steel blocks. When different sizes are required, the toolmaker rubs a combination of these blocks together to create the desired size (Fig. 3.3). The blocks stay together because the surfaces are smooth enough to allow the steel molecules at the surface to interact.

Chemical Absorption. When the chemical component of a bonding medium penetrates a substrate, there can be a transference of surface molecules between the substrates (Fig. 3.4). After molecules migrate to the realm of the other substrate, the solvent component of the bonding medium freezes the transferred molecules, rendering the bond. The chemical absorption mechanism does not fall in the direct adhesive bonding classification. It is often referred to as "solvent bonding."

A hobbyist building a model airplane utilizes the solvent bonding technique. The model glue is actually a solvent that softens and solvates the

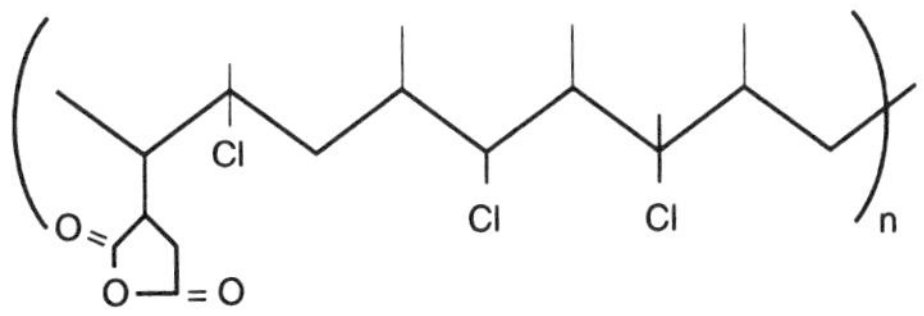

Fig. 3.2 Chlorinated polyolefin (CPO). Source: R.J. Clements, G.N. Batts, J.E. Lawniczak, K.P. Middleton, C. Sass, How Do Chlorinated Poly(olefins) Promote Adhesion of Coatings to Poly(propylene), *Prog. Org. Coatings,* Vol 24, 1994, p 43–54

32 / Decoration and Assembly of Plastics

plastic model components. As the parts are held together, the molecular transference takes place. After a few moments, the solvent flashes off and the transferred molecules are locked in place, creating the bonded assembly.

Chemical (Covalent) Bonds. Chemically compatible adhesives and adherends allow covalent bonding across the interface. Because there is a true chemical bond, the strength of the adhesive bond is strong and durable. This form of bonding is the typical mechanism for many adhesive systems used today.

Electrostatic attraction is the most common adhesion mechanism. Electrostatic charges can cause dust and particulates to be attracted to an oppositely charged surface. This phenomenon can be seen everyday if you wipe a television screen. The electrostatic attraction of polar groups in the adhesive and adherends provide most of the cohesive strength of the bond. The polar

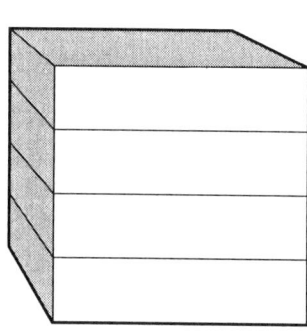

Fig. 3.3 Stacked Joe Blocks made of highly polished steel that stay together due to their close proximity

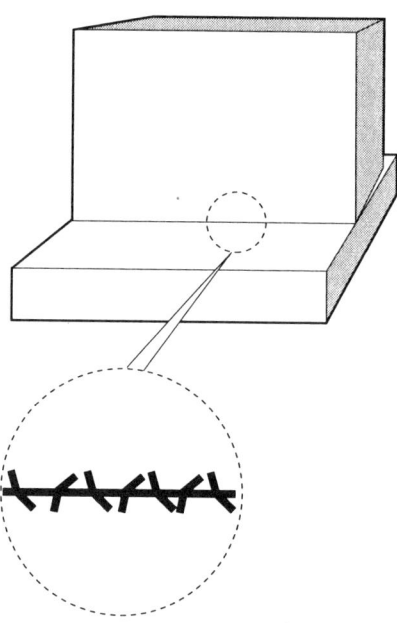

Fig. 3.4 Diffusion model. Molecules made flexible by solvent are allowed to migrate across substrate interface. When solvent flashes off, molecules return to rigid state spanning the interface, thus creating the bond.

forces also account for polymer entanglement, which provides cohesive strength of the adhesive polymer.

Types of Adhesives

There are two main categories of adhesives based on the markets and end-use applications of these products.

Industrial adhesives have maximum properties, strength, and durability for use in industrial applications. They may be challenging to use and require special storage, handling/mixing, and application technique. Additionally, the industrial adhesive market segment may be expensive.

Consumer adhesives are those adhesive products available to the public and available in department stores and hobby outlets. Consumer adhesives may be lower in strength, properties, and durability. To facilitate mixing multicomponent systems, such as epoxy, some consumer adhesives are diluted with fillers (inert additives) to facilitate mixing proportions.

Function of Adhesives

Fastening. Adhesives are able to transmit loads from one member of a joint to another versus mechanical fastening.

Sealing and insulation may be secondary benefits of using an adhesive. Sealing becomes very important when comparing (metallic) welding of metal-to-metal products versus (polymer) adhesive bonding of metal-to-metal products.

Resistance to Corrosion and Vibration. Because the polymer adhesive is electrically dielectric, there is no galvanic action, such as occurs with metal-to-metal bonds. Adhesives tend to dampen vibrations by dispersing loads.

Resistance to Fatigue. Vibration dampening also results in improved fatigue resistance.

Adhesion Selection Process

Selecting an adhesive is similar to selecting a plastic material for a specific application. There needs to be a thorough assessment of all aspects of the application and the environment. Key steps should include:

1. Determine strength and durability requirements of the application: humidity, thermal shock, temperature exposure, solvent exposure
2. Understand the substrate(s)
   a. Compatibility: is the adhesive compatible with the substrate(s)?
   b. Similarities: metal to metal, plastic to plastic
   c. Dissimilarities: metal to plastic, plastic to different plastic
   d. Cleanliness: can surface conditions be met?
   e. Migratables: does substrate contain components that move to the surface over time and affect the bond?
   f. Degradation: will the structure and strength of the bond degrade?
3. Understand the loading environment: short term, long term, cyclical, and magnitude
4. Understand assembly requirements: surface preparation, priming, fixturing/dispensing
5. Consider recyclability and rework issues

Adhesive Material Selection Matrix

In order to simplify the material selection process, many product designers utilize a material selection matrix, such as the one shown in Fig. 3.5. The matrix allows a direct comparison between the desired end-use properties and the actual properties available from the candidate materials. Additionally, the matrix assigns value and sorts these properties to aid in the final selection. The material matrix has several basic steps:

1. Identify as many material properties/attributes as possible and the demands of the application.
2. Rate the properties/attributes with a weighted value (9 = critical, 6 = desirable, 3 = optional).
3. List the candidate materials.
4. Rank the materials, relative to each other, in each property/attribute category. For example, if there are 4 materials being compared, the material that best meets the property/attribute receives a 4. The material that least meets the property/attribute receives a 1.
5. Multiply the rating (step 2) by the ranking (step 4).
6. Add the products of step 5.

The material with the highest sum is the top candidate. This matrix analysis tool can be readily adapted to different processes for optimizing the selection process. Figure 3.6 shows an example of a completed matrix.

Classification of Adhesives

Adhesive classification can be based on function type and chemical family.

Function Type

Consumer adhesives have been designed for low cost and ease of application. These adhesives usually have a lower bond strength and limited compatibility with several substrates.

Industrial adhesives are high in bond strength and substrate compatibility. Application and dispensing may require specialized equipment.

Fig. 3.5 Blank materials matrix

36 / Decoration and Assembly of Plastics

Structural adhesives, a segment of the industrial adhesive market, have high bond strength that is often comparable to welding with both metal and plastic substrates. Many structural adhesives are further categorized by their setting or curing temperature range:

Setting	Temperature, °C
Cold	<20
Room temperature	20–30
Intermediate	31–99
Hot	>100
Pressure sensitive	20–30

Rating values					
9 = Critical					
6 = Desirable					
3 = Optional					
Desired properties	Rating	Adhesive 1 Cyanoacrylate	Adhesive 2 Epoxy A	Adhesive 3 Epoxy B	
Bond Strength (Tensile)	9	2	1	3	Rank
		18	9	27	RatingXRank
Cost	6	3	1	2	Rank
		18	6	12	RatingXRank
Appearance	3	3	1	2	Rank
		9	3	6	RatingXRank
Compatibility	6	2	1	3	Rank
		12	6	18	RatingXRank
					Rank
					RatingXRank
					Rank
					RatingXRank
					Rank
					RatingXRank
					Rank
					RatingXRank
TOTAL (Sum of products)		57	24	63	

Fig. 3.6 Completed materials matrix

Hot melt adhesives utilize thermoplastic materials that bond as a function of their melt phase compatibility with substrates. Hot melt adhesives are common to household applications and hobbyists. The hot glue guns available today are most often a variation of low density polyethylene. Hot melt adhesives for the industrial markets are most often polyamide (nylon) materials.

Pressure sensitive adhesives (PSA) include labels, stickers, and tapes.

Chemical Families

Silicones, epoxies, cyanoacrylates, and hot melt are several of the chemical families of plastics. Under the heading of chemical families there can be additional categorization, such as natural or synthetic polymer adhesives. A natural adhesive requires little or no additional chemical modification to be used. Some natural gum materials fall into this category. Polymer adhesive systems (the most prevalent form) begin with organic material and are chemically polymerized to produce the final product.

Thermoset or Thermoplastic

Thermoset adhesives are those systems that form crosslinks across the long chain polymers, which may also be referred to as "curing." The crosslinks render the cured polymer infusible, and this operation is irreversible. Therefore, the bond is prevented from softening at higher temperatures, and its chemical resistance is significantly improved. Thermosetting adhesives have a defined shelf life and pot life. Shelf life is the time the adhesive components can be stored (separately) and still be able to produce a quality bond. Pot life is the duration of time the adhesive system can be used or applied after mixing has occurred. Examples of thermosetting adhesives include epoxy (Fig. 3.7) and silicones (Fig. 3.8).

Thermoplastic adhesive systems are also long chain molecules (Fig 3.9). However, these long molecules do not have any crosslinking. They are susceptible to softening at higher temperatures and have much lower chemical and environmental resistance than thermosetting adhesives. Thermoplastic adhesives, like thermosetting adhesives, have a defined shelf life. Only the multicomponent systems have a defined pot life. Examples of thermoplastic adhesives include hot melt systems, polyvinyl acetate (white glue), polyvinyl alcohol (used on stamps), and cyanoacrylates.

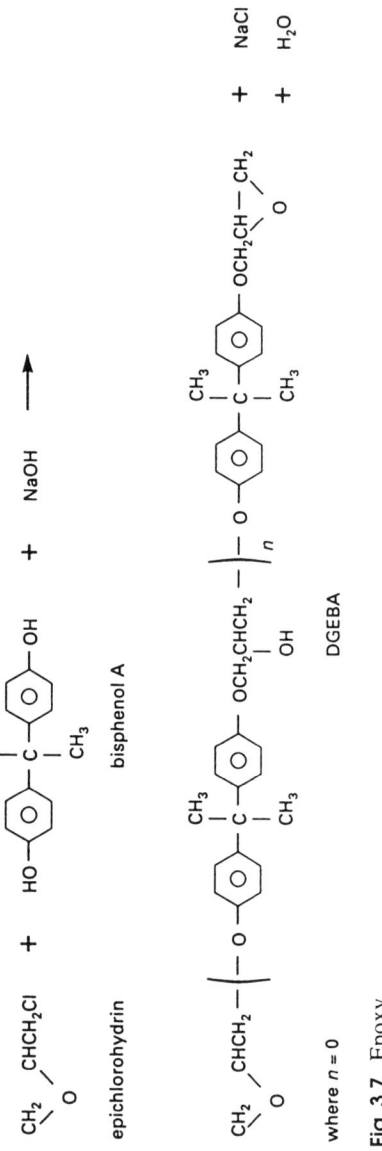

Fig. 3.7 Epoxy

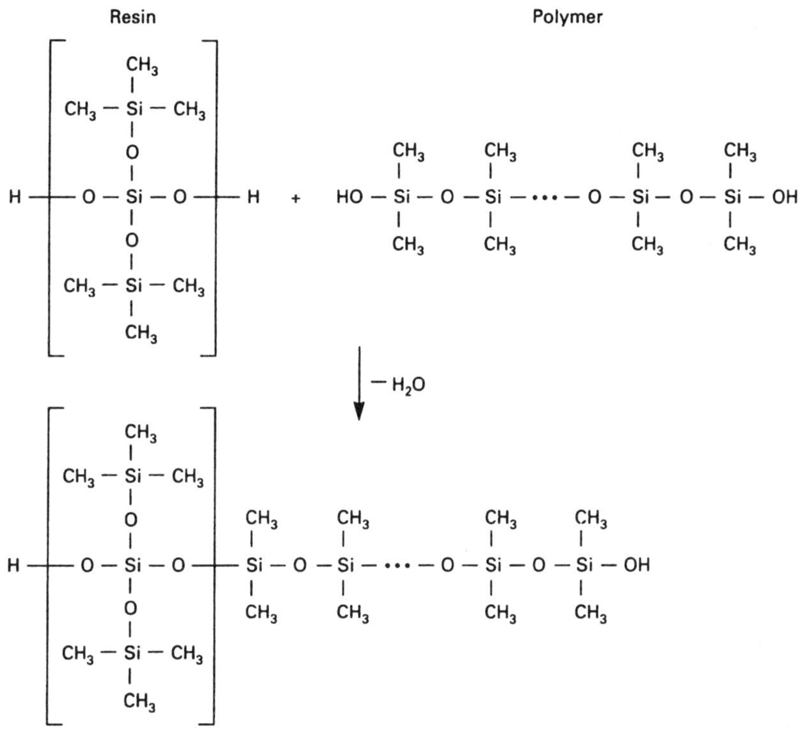

Adhesives / 39

Fig. 3.8

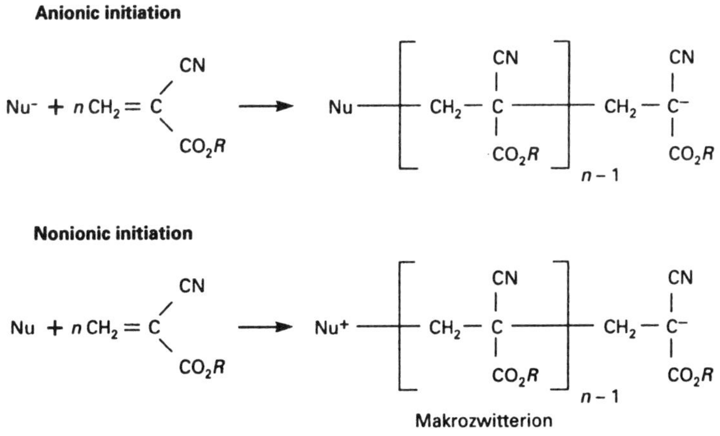

Fig. 3.9 Examples of thermoplastics

Physical Form

One-Component System

One-component systems are available in both thermoplastic and thermosetting systems. They are easier to use, because there is no mixing required. Some one-component systems include white glue, 2-cyanoacrylate (super glue), thread fastening systems, and hot melt systems.

Two-Component or Multicomponent System

Multicomponent adhesive systems are, most often, thermosetting. They require precise mixing ratios and thorough mixing for a good bond. Some multicomponent systems include epoxies, some polyurethanes, and silicone-based materials. Multicomponent thermoset adhesives systems may be further categorized by the stages of their cure.

A Stage. Systems are cured prior to mixing the components. The polymer is still soluble, and it is often liquid or a viscous liquid.

B Stage. After mixing, however, crosslinking has begun. The polymer is still soluble, and it is often liquid or a viscous liquid.

C Stage. The polymer adhesive system is crosslinked (cured) and is insoluble and solid (or unable to be applied).

Film

Film format adhesives are used to bond broad substrate surfaces. The adhesive can be in a one-component film or as a multicomponent film. The multicomponent systems are already mixed and compounded and are often incorporated with a reinforcement carrier, such as glass cloth. Most often, film-based adhesives are utilized in the composite construction of aerospace systems and the automotive industry.

Assessing Adhesive Properties

Adhesive properties will vary with the adhesive, the substrates being used, and the loading environment. Certainly, there are other factors that may affect bond strength, but these three are the most significant.

Adhesive

The analysis of variation in bond strength versus adhesive type must consider several factors.

Adhesive chemistry includes computability with substrates as well as the nature of the polymer. In other words, is it thermoset or thermoplastic in nature?

Adhesive Physical Form. Is the adhesive system liquid, paste, or film? Is the adhesive multicomponent, requiring mixing? Is the viscosity of the adhesive low enough to promote wetting of the substrate?

Strength Orientation. Does the bond strength of the adhesive tend to be isotropic (strength equal in all directions) or anisotropic (strength not equal in all directions)? A good example (although somewhat risky to perform) is the consumer form of the adhesive 2 cyanoacrylate ("super glue"). Advertised as having superior bond strength, the advertisements fail to state that the bond is significantly stronger in tension than in shear. If the consumer mistakenly bonds thumb and forefinger together and attempts to pull them apart, there is risk of torn skin and a lot of pain. If, however, the same fingers are bonded and the consumer rolls the two digits apart, the bond will give way much easier.

Substrate Chemistry. What is to be bonded has a great affect on how well it will be bonded. As mentioned earlier, the surface chemistry of the substrate needs to be understood relative to the chemistry of the adhesive. Glass bonds differently than plastic, and rubber binds differently than paper. Most adhesive manufacturers have bonding charts to facilitate the adhesion assessment process.

Substrate Form. In addition to the chemistry of the substrate, the physical condition must be considered. Simply stated, surfaces might be too rough to allow proper wetting, or too smooth to effect a good bond.

Loading Environment

Adhesive loading is usually defined in four distinct ways (Fig. 3.10):

- *Tensile:* The force attempting to pull the substrates apart is normal to the adhesive and substrates.
- *Shear:* The force attempting to pull the substrates apart is parallel to the substrates and adhesive.

- *Peel:* The force attempting to pull the substrates apart is parallel on one substrate, and the other substrate is pulled back from the front and worked parallel in the opposite direction.
- *Cleavage:* Both substrates are forced back from the same starting area.

Bond Strengths

It is often the objectives of assembly operations to have the strength of the bonded area greater than that of the substrates being bonded. Plastics are no different. Plastic substrates, however, can vary greatly in their flexural properties (rigidity). Figure 3.11 shows what may occur when two plastic substrates are bonded utilizing a lap joint. When the substrates are pulled in what initially appears to be a tensile mode, it is truly a shear mode. Depending on the nature of the substrates, it is possible for the bond to be strong enough to hold while the substrates distort. If the substrates do not immediately fail, they could twist, resulting in a load conversion. The original shear loading (Fig. 3.11a) can be altered to become more of a tensile load (Fig. 3.11d). Depending on the original load analysis and the properties desired in the materials selection matrix, the failure mode could be a surprise to the design engineer.

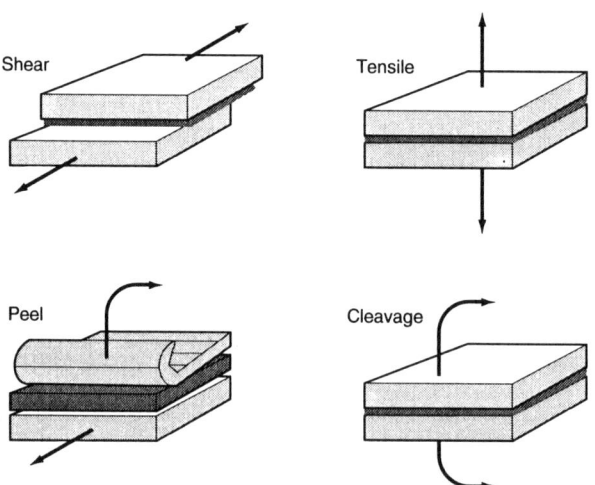

Fig. 3.10 Adhesive-substrate loading scenarios

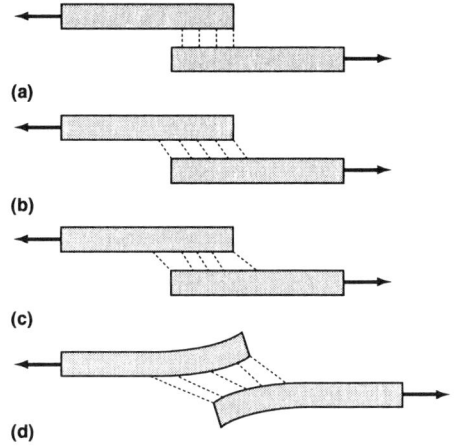

Fig. 3.11 Adhesive-bonded lap shear samples

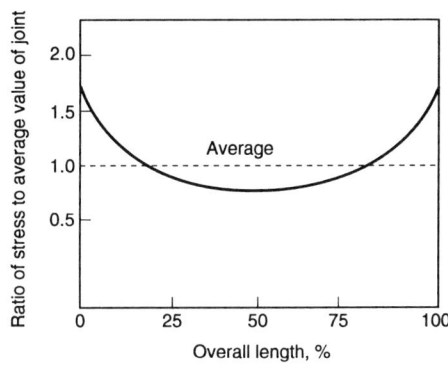

Fig. 3.12 Stress concentrations near ends of overlap joints

The distortion of the substrate in failure tends to place a high stress concentration level at the ends of the overlap. The stress ratio levels at the plastic overlap joint are much higher than those of other materials, such as metal. Therefore, the plastic substrate fails at the overlap. Figure 3.12 shows how this stress concentration can vary across the overlap joint of the bonded substrates.

4

Welding Assembly of Plastics

Spin Welding

Spin welding is a plastics assembly operation that requires parts to be manufactured from thermoplastic materials. The essence of the process (Fig. 4.1) involves creating a frictional heat via high speed rotation (5000 rpm). Either counter-rotating plastic parts, one fixed and one rotating part, or a fixed part and a rotating mandrel can create the rotational friction (Fig. 4.2).

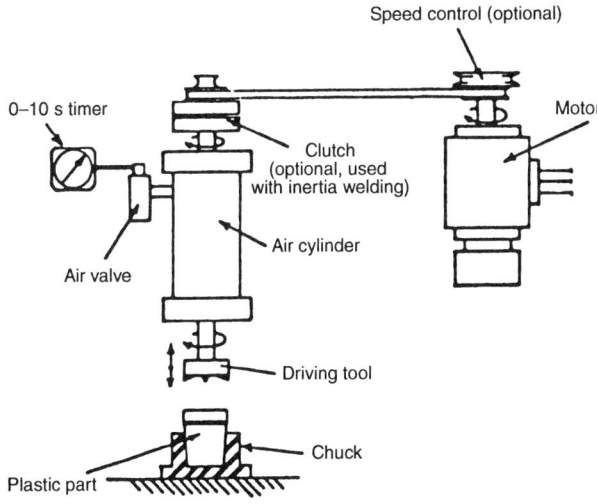

Fig. 4.1 Basic equipment requirements for spin welding

The friction, in turn, creates a thin film of plastic melt at the interface of the two parts. At a rotational speed of 5000 rpm, this film is created rather quickly. After the creation of the melt film, the rotation of the part(s) is abruptly stopped. An air cylinder forces the two parts together until the melt film solidifies and the assembly is made.

Material Compatibility. To affect an acceptable bond between two thermoplastic materials, it is preferable to have the parts made with the same plastic. If this accommodation is not possible, the two materials should meet these basic criteria:

- *Compatibility:* There should be no adverse reaction to the materials blending in each domain, which may include additives, such as color/colorants.
- *Similar melting ranges:* In order to have the plastic molecules flow into each domain, the melt temperature ranges must be similar. Most plastics melt across a range of temperatures versus a specific melting point. If the plastics vary significantly in the melt ranges, it is likely that the plastic with the lower melt range will soften, and the material with the higher melt range will remain solid. Thus, the welding process would be ineffective.

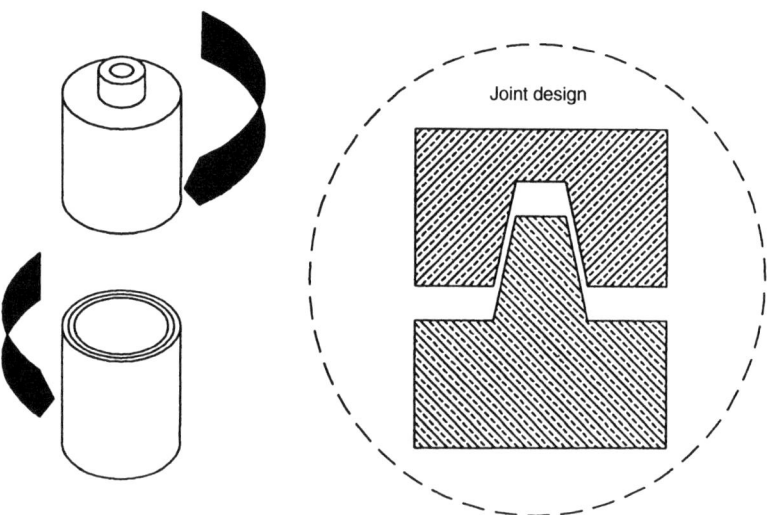

Fig. 4.2 Counter-rotating plastic parts for spin welding—a simple, low-cost assembly technique

- *Similar flow rates:* Although materials may meet the similar melt range criteria, it is possible that the flow rates may be significantly different and, therefore, render the welding process ineffective. Flow rates of plastics relates to how far the plastic travels when it encounters a specific force and temperature. Plastics with similar melt ranges may not have similar flow rates. When welding is attempted, the plastic with the greatest flow rate may flow away from the welding area prior to the actual melting of the second plastic part.

Applications. Typical applications for spin-welded products include fuel filters and beverage coolers. Two halves of a fuel filter may be made of chemical resistant plastic. Prior to assembly, a paper filter is inserted in one of the halves. When assembled, the sealed plastic filter contains the irreplaceable filter and is leakproof. Modern round beverage coolers utilize a hollow-walled system between the plastic parts that can be filled either with air or insulating foam. The parts are then assembled using the spin welding process to seal them with an insulating area in between.

Advantages and Disadvantages. The speed of the spin welding cycle is a function of the size of the part and the plastic materials being assembled; however, a 1 to 2 s cycle is typical. Another advantage is that no third material components, such as adhesives, are required to create the assembly bond. Therefore, the assembly is more readily recycled, and the assembly of plastic material cannot be bonded with adhesives. Many designers appreciate that spin welding can also be used to eliminate assembly hardware and improve the design for assembly aspects of the product. Molded-in bosses can be formed into a rivet-like shape (Fig. 4.3) in order to fasten components together.

Joint area design for spin-welded plastic products must allow for the proper bearing surface between the parts to affect the bond. Also, when the two parts are forced together under pressure, there may be excess plastic melt that is forced outside of the joint area, which may be unsightly and demand a secondary trim operation. The spin weld joint designs shown in Fig. 4.4 demonstrate how designers incorporate a trap or space for this unwanted flash.

Fusion Bonding

Fusion bonding is another name for what is commonly referred to as "hot plate welding." The process (Fig. 4.5) utilizes a heated plate, usually made of aluminum, which is sandwiched between the plastic parts to be assembled. The hot plate, which may also have a protective covering, is placed within 0.254 mm (0.010 in.) of the two surfaces to be bonded. When the plastic surfaces are softened to an adequate level, the heated plate is quickly removed, and the two parts are forced together. The edges, where the hot plate has softened the plastic, are allowed to have plastic material flow across the interface.

Material Compatibility. As with the spin welding process, the plastic materials to be assembled utilizing the fusion bonding process must exhibit these three characteristics: basic compatibility (no adverse effects when assembled), close melting temperatures (within 14 °C, or 25 °F), and similar flow characteristics.

Applications. Fusion bonding is used to assemble large, relatively flat designs. Some automotive component manufacturers prefer to assemble tail light assemblies (red, clear, and amber parts) using fusion bonding instead of the more expensive multicolor injection molding process. Fusion bonding equipment is very inexpensive (compared to injection molding), and tooling is very versatile and low in cost.

Advantages and Disadvantages. The speed of the fusion bonding cycle (5 to 20 s) is longer than other assembly processes; however, it is often used

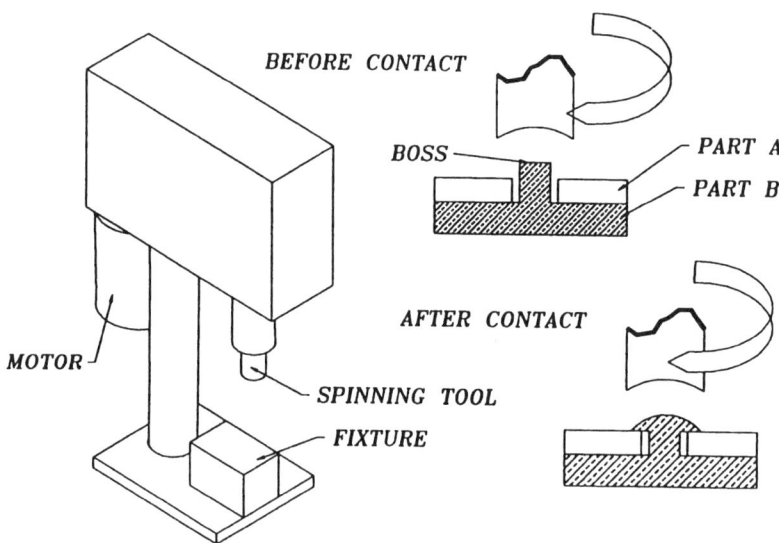

Fig. 4.3 Molded-in bosses can be formed into a rivet-like shape

for larger, more complex products. Another advantage is that no third material component, such as adhesives, is required to create the assembly bond. Therefore, the assembly is more readily recycled, and the assembly of plastic material cannot be bonded with adhesives.

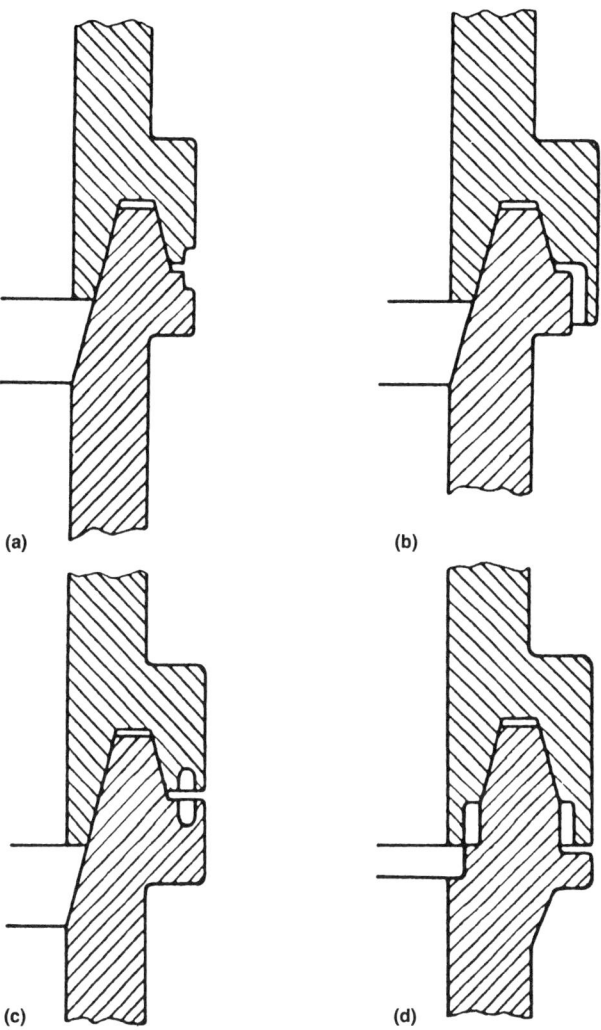

Fig. 4.4 Weld joints with flash traps suitable for (a) long cylindrical parts, (b) various gas containers where a bead cannot be left, and (c) and (d) designs that hide or trap excess weld flash

Joint Area Design. Fusion bonding does not require complex joint design, because this assembly process demands flat bearing surfaces. Product designs are encouraged to provide a distinct land area around the bond area.

Vibration Welding

Vibration welding is a plastic assembly process that utilizes the frictional heat generated by vibrating or sliding two parts against each other (Fig. 4.6). The actual displacement of the sliding parts is relatively low (3.18 to 6.35 mm, or 0.125 to 0.250 in.), and the frequency of vibration is relatively small (120 to 240 Hz). The friction resulting from the vibration creates a melt film that is used to bond the parts together. Once the melt film is created, the vibrations are stopped, and the parts are quickly positioned to the proper location relative to each other. At that time, pressure is applied to the parts to force them together. The plastic melts where held until the melt cools, and the bond is created.

Material Compatibility. As with the spin welding and fusion bonding processes, the plastic materials to be assembled utilizing the vibration welding process must exhibit these three characteristics: basic compatibility (no adverse effects when assembled), close melting temperatures (within 14 °C, or 25 °F), and similar flow characteristics.

Applications. The vibration welding process lends itself to large, flat product designs. It is important to note that the vibration motion can either be linear or angular (Fig. 4.6).

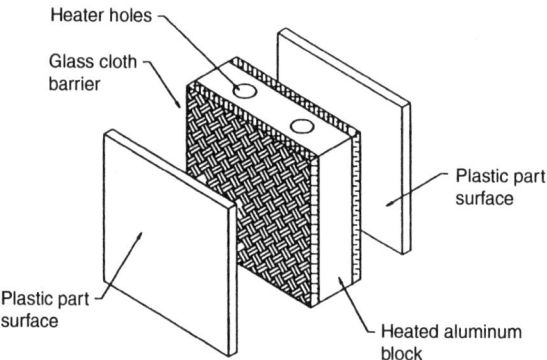

Fig. 4.5 Fusion bonding equipment

Advantages and Disadvantages. The speed of the fusion bonding cycle (5 to 20 s) is longer than other assembly processes; however, it is often used for larger, more complex products. Another advantage is that no third material component, such as adhesives, is required to create the assembly bond. Therefore, the assembly is more readily recycled, and the assembly of plastic material cannot be bonded with adhesives.

Joint Area Design. Several vibration weld joint designs are shown in Fig. 4.7. As with the spin welding assembly process, it is important to recognize that there could be a flash bead generated. The joint design can provide the necessary space to house the flash and prevent an unsightly bead.

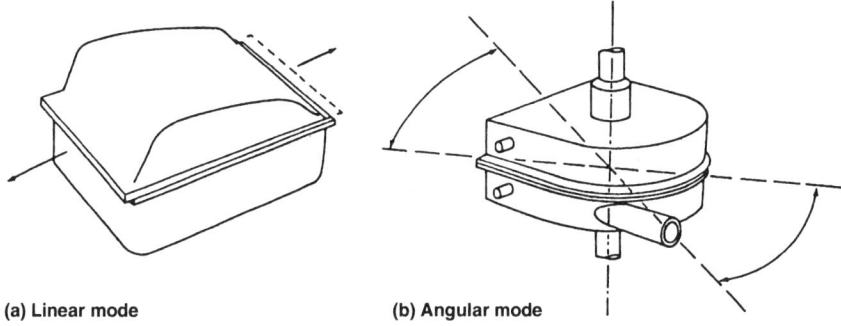

(a) Linear mode   (b) Angular mode

Fig. 4.6 Principle of vibration welding

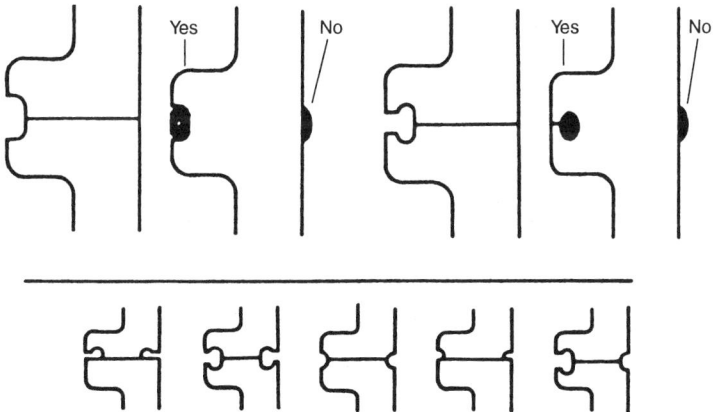

Fig. 4.7 Basic flash trap design and variations

Ultrasonic Welding

Ultrasonic welding, although not unique to plastics, is used extensively throughout the plastics industry. The term ultrasound means "beyond hearing." The human ear can sense sound vibrations to about 18 kHz. The frequency of ultrasonic vibrations used for plastic welding is usually 20 to 40 kHz. Ultrasonic-related equipment is used in many areas: ultrasound is used to assess the development of the fetus while still in the womb; ultrasonic degreasing is used to facilitate the cleaning of detailed components; and ultrasonic testing is used to assess engine blocks and large castings for defects, such as voids.

Ultrasonic welders (Fig. 4.8, 4.9) consist of four major components:

- *Power supply:* Controls the power to the converter by receiving electric power at 60 Hz and amplifying it to 20 to 40 kHz
- *Converter:* Utilizes a piezoelectric element to convert the 20 to 40 kHz electric input to a 20 to 40 kHz mechanical vibration output
- *Booster:* Increases (or decreases) the amplitude of the mechanical vibration
- *Horn:* Focuses the mechanical vibration to the tip of the horn where it contacts the plastic part or an insert

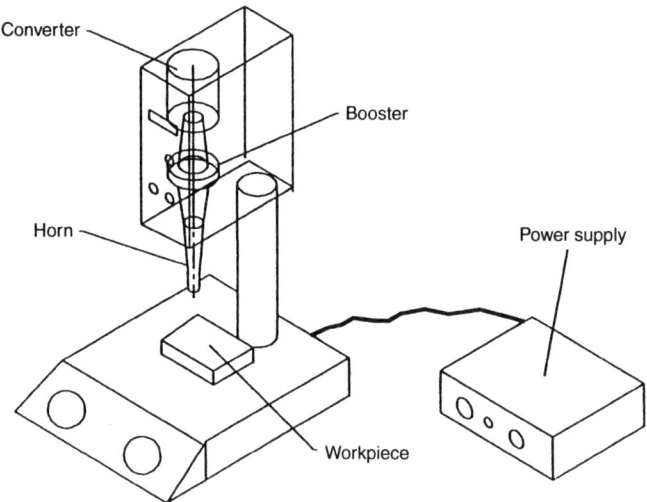

Fig. 4.8 Ultrasonic welder components

Many individuals, when first introduced to the ultrasonic welding process, believe that the entire horn unit moves up and down like a jackhammer or that sound is somehow involved in the welding of the plastic.

In an overview of the ultrasonic welding process, the mechanical vibrations at the tip of the horn do the work (Fig. 4.10). The tip or end of the horn has a displacement of about 1 to 1.5 mm (0.003 to 0.005 in.). The frequency of this displacement is the frequency rating of the welder (20 to 40 kHz).

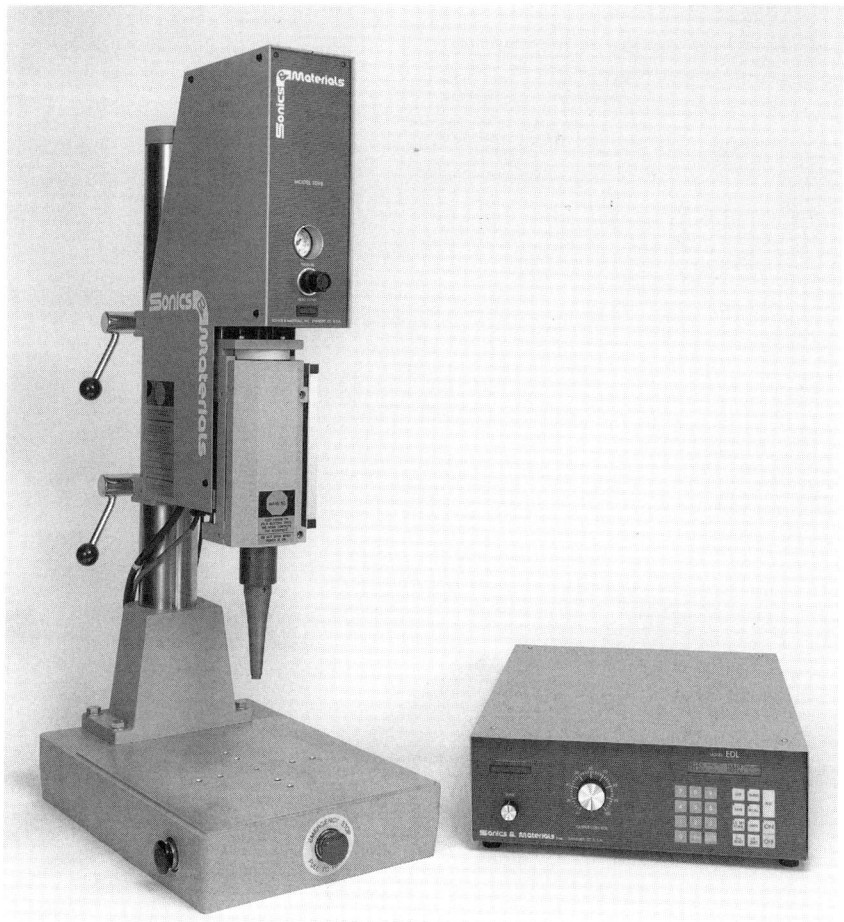

Fig. 4.9 Ultrasonic welding equipment. Courtesy of Sonics and Materials Inc.

54 / Decoration and Assembly of Plastics

The basic principle of ultrasonic assembly involves conversion of high-frequency electrical energy to high-frequency mechanical energy in the form of reciprocating vertical motion, which, when applied to a thermoplastic, can generate frictional heat at the plastic/plastic or plastic/metal interface. In ultrasonic welding, this frictional heat melts the plastic, allowing the two surfaces to fuse together. In ultrasonic staking, forming, or insertion, the controlled flow of the molten plastic is used to capture or retain another component in place (staking/forming) or encapsulate a metal insert (insertion).

Thermoplastics can be ultrasonically assembled, because they melt within a specific temperature range, whereas thermosetting materials, which degrade when heated, are unsuitable for ultrasonic assembly. Weldability of any thermoplastic depends on its stiffness or modulus of elasticity, density, coefficient of friction, thermal conductivity, specific heat, and melt temperature (T_m) or glass transition temperature (T_g).

Rigid plastics exhibit excellent welding properties, because they readily transmit vibratory energy. Soft plastics, having a low modulus of elasticity, attenuate the ultrasonic vibrations and, as such, are more difficult to weld. In staking, forming, or spot welding, the opposite is true. Generally the softer the plastic is, the easier it is to stake, form, or spot weld.

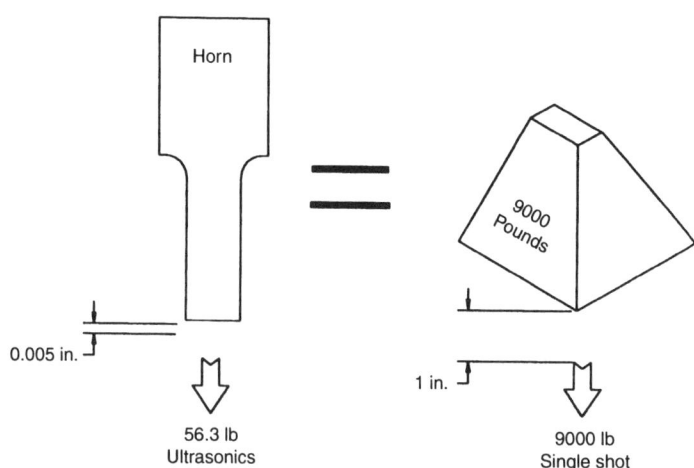

Fig. 4.10 Force equivalent to ultrasonic vibrations

Plastic Materials and Ultrasonics

Resins are classified as amorphous or crystalline. Ultrasonic energy is easily transmitted through amorphous resins, and these resins lend themselves readily to ultrasonic welding. Amorphous resins are characterized by random molecular arrangements and a broad melting range, which allow the material to soften gradually before melting and to flow without prematurely solidifying. Because the molecular structure in the crystalline resins attenuate a great amount of energy, crystalline resins do not readily transmit ultrasonic energy. They require higher energy levels than amorphous resins. A high, sharply defined melting point that causes melting and resolidification to occur rapidly characterizes these resins. For these reasons, when welding crystalline resins, higher amplitude and energy levels should be used and special consideration given to joint design.

Ultrasonic Welding Characteristics

Before discussing welding characteristics, the difference between near-field and far-field welding must be understood. Near-field welding refers to welding a joint located 6 mm (¼ in.) or less from the area of horn contact, while far-field welding refers to welding a joint located more than 6 mm (¼ in.) from the horn contact area. The greater the distance is from the point of horn contact to the joint, the more difficult it will be for the vibration to travel through the material and for the welding process to take place.

The differential, if any, in the melt temperature of the materials being welded should not exceed 17 °C (30 °F), and the molecular structure of the materials should be compatible (i.e., blends, alloys, copolymers, and terpolymers).

Moisture content, mold release agents, lubricants, plasticizers, fillers, reinforcing agents, regrinds, pigments, flame retardants, and resin grade are all variables that can influence weldability.

The moisture content of parts molded from resins that are hygroscopic (moisture absorbent) can be problematic. Nylon (and to a lesser degree polycarbonate and polysulfone) presents most of the problem, and parts molded in hygroscopic resins should be stored in sealed polyethylene bags with an appropriate desiccant immediately after molding. If moist parts are welded, the escaping vapors may cause voids and fissures in the molten material resulting in a weld of poor integrity.

Mold release agents, such as zinc stearate, aluminum stearate, fluorocarbons, and silicones, are not compatible with ultrasonic welding. If it is necessary to use a mold release agent, the paintable/printable grades that permit painting and silk screening should be considered. Other release agents should be removed with either Freon TF for crystalline resins or a 50/50 solution of water and liquid detergent.

Lubricants, whether waxes, stearates, or fatty esters, reduce intermolecular friction within the polymer and inhibit the ultrasonic assembly process. However, because they are generally dispersed internally, the effect is usually negligible.

Plasticizers, which usually impart flexibility and softness to a resin, can interfere with the ability of a resin to transmit vibratory energy. Plasticizers that are FDA-approved do not present as much of a problem as metallic plasticizers, but experimentation is recommended.

Although fillers and reinforcing agents, such as glass and talc, can considerably increase the ultrasonic weldability of a thermoplastic, they should be judiciously used. When additive content exceeds 10%, premature horn wear may result, and specially treated steel or carbide-faced titanium horns might be required. When filler content approaches 35%, there may be insufficient resin at the surface to obtain hermetic seals. When filler content exceeds 40%, insufficient plastic is present at the interface to form a positive bond. A reinforcement composed of long glass fibers is always more problematic than a reinforcement composed of short glass fibers.

Regrind. Ultrasonic assembly is one of the few methods that permits regrinding of assembled parts, because no foreign substance is introduced into the resin/parts. Ultrasonic assembled parts manufactured from parts made from regrind present no problem, providing that the percentage of regrind is not excessive, and the plastic has not been degraded. Regrind limitations suggested by the resin suppliers should be observed.

Pigments. Although most pigments do not interfere with the ultrasonic assembly process, some oil-based colorants can adversely influence weldability. Non-oil based pigments should be used.

Flame retardants greatly affect the weldability of thermoplastics and the effects of the various additives should be investigated experimentally prior to resin selection.

The grade of resin can have a significant influence on weldability. There is a great difference between injection/extrusion grades and cast grades. Molecular weight, melt temperature, and modulus of elasticity are quite different. Injection/extrusion grades should only be used with injection/extrusion grades, and cast grades should only be used with cast grades.

Table 4.1 lists the characteristics of thermoplastics, and Fig. 4.11 shows the compatibility of thermoplastics.

The basic principle of ultrasonic assembly involves conversion of high frequency electrical energy to high frequency mechanical energy in the form of reciprocating longitudinal motion that, when applied to a thermoplastic, generates frictional heat at the plastic/plastic or plastic/metal interface, creating a localized melt.

Table 4.1 Characteristics of thermoplastics

Material	Spot welding	Staking/swaging	Inserting	Welding field Near	Welding field Far
Amorphous					
ABS	E	E	E	E	G
ABS/polycarbonate	G	G	G	G	F
ABS/PVC	G	G	F	G	F
Acrylic	G	F	G	G	F
Acrylic multipolymer XT	G	G	G	G	F
Acrylic/PVC	G	G	F	G	F
Acrylic, impact modified	F	F	P	F	P
Butadiene, styrene (BDS)	G	G	G	G	F
Cellulosics: CA, CAB, CAP	P	G	E	P	...
Modified phenylene oxide	E	E	E	E	G
Polyarylate	F	F	G	G	F
Polycarbonate	G	F	G	G	F
Polyetherimide	G	G	E	E	G
Polystyrene, G.P.	F	F	G	E	E
Polystyrene, impact modified	F	F	G	G	P
PVC, rigid	F	G	E	P	P
PVC, flexible	P	...	...	P	...
SAN, NAS, ASA	F	F	G	E	E
Styrene maleic anhydride	E	E	E	E	G
Sulfone polymers	F	F	G	G	F
Crystalline					
Acetal copolymer	F	F	G	G	F
Acetal homopolymer	F	F	G	G	F
Fluoropolymers	...	...	...	P	...
Nylon	F	F	G	G	F
PC, PET	G	G	E	E	G
Polyester, PBT	F	F	G	G	F
Polyester, PET	F	F	G	G	F
Polyetheretherketone	G	G	E	E	G
Polyethylene (LDPE, HDPE)	G	F	G	P	P
Polyethylene (UHMW)	...	...	...	...	...
Polymethylpentene	G	F	E	F	P
Polyphenylene sulfide	F	P	G	G	F
Polypropylene	E	E	G	F–P	P

E, excellent; G, good; F, fair; P, poor. All others are unsuitable for ultrasonic assembly.

Ultrasonic Staking

In ultrasonic staking, also referred to as ultrasonic "heading or riveting," the controlled flow of the molten plastic is used to capture or retain another component, usually of a different material. Ultrasonic staking provides an alternative to welding when the two parts consist either of dissimilar materials that cannot be welded or simple mechanical retention of one part relative to another is adequate (as distinct from molecular bonding).

The most usual application involves the attachment of metal to plastic. A hole in the metal part receives a premolded plastic boss. The horn tip, vibrating at high frequency, contacts the boss and, through friction, creates

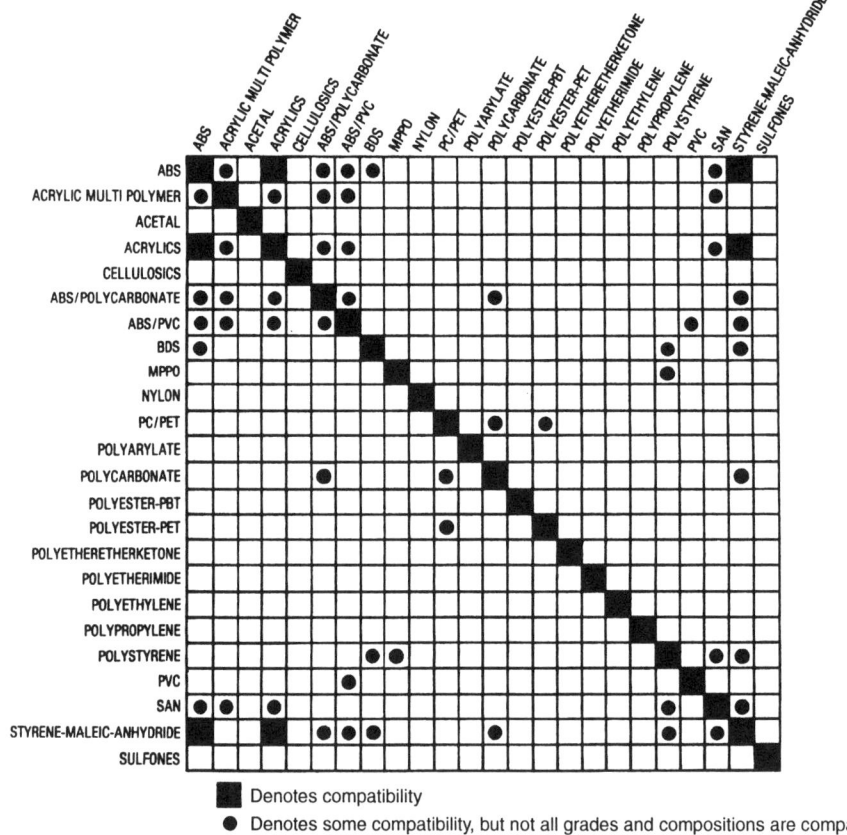

Fig. 4.11 Compatibility of thermoplastics

localized heat. As the boss melts, the light pressure from the horn forms a head to a shape determined by the horn tip configuration. When the vibrations stop, the plastic material solidifies, and the dissimilar materials are fastened together.

Unlike ultrasonic plastics welding, staking requires that out-of-phase vibrations be generated between the horn and the plastic surfaces. Light, initial contact pressure is therefore a requirement for out-of-phase vibratory activity within the limited contact area. It is the progressive melting of the plastic boss under continuous, but light, pressure that forms the head. When staking, low pressure rather than high pressure is usually recommended.

With staking, tight assemblies are possible because mating parts are clamped under the pressure of the horn until the rivet head solidifies. There is no elastic recovery as is the situation with heat staking or cold forming.

The decision to use ultrasonic staking should be considered when the parts to be assembled are still in the design stage. Several configurations for boss/cavity design are available, each with specific features and advantages. The selection is determined by type of plastic, part geometry, assembly requirements, machining and molding capabilities, and cosmetic appearance. The principle of staking is the same for each: the area of initial contact between the horn and the boss is kept to a minimum in order to concentrate the energy and produce a rapid melt.

The integrity of an ultrasonically staked assembly depends greatly upon the geometric relationship between the boss and the horn cavity. Proper design will produce optimal strength with minimum flash. Whenever possible, the bosses should be designed with an undercut radius at the base to prevent fracturing or melting. Holes in the mating parts should be radiused or at least deburred. Long bosses should be avoided, and properly designed bosses taper from the base to the top. The boss should be properly located and rigidly supported from below to ensure that the energy will be dissipated at the horn/boss interface, rather than exciting the entire plastic assembly and fixture.

Best staking results are obtained when the ultrasonic vibrations are started before the horn contacts the boss. Thus, cold forming is prevented, and gradual reforming of the boss is allowed. The pretriggering of the ultrasonic vibrations is normally accomplished using a pretrigger switch located on the welder itself. To obtain repeatable results when staking, the distance that the horn travels should be consistent and limited by the positive stop adjustment.

60 / Decoration and Assembly of Plastics

The standard flared stake (Fig. 4.12) satisfies the requirements of most applications. This stake is recommended for bosses with an outside diameter (OD) of 1.6 mm ($\frac{1}{16}$ in.) or larger, and it is ideally suited for low density, nonabrasive, amorphous plastics.

The spherical stake (Fig. 4.13) is preferred for bosses with an OD less than 1.6 mm ($\frac{1}{16}$ in.) and is recommended for rigid crystalline plastics with sharp highly defined melting temperatures, for plastics with abrasive fillers, and for materials that degrade easily.

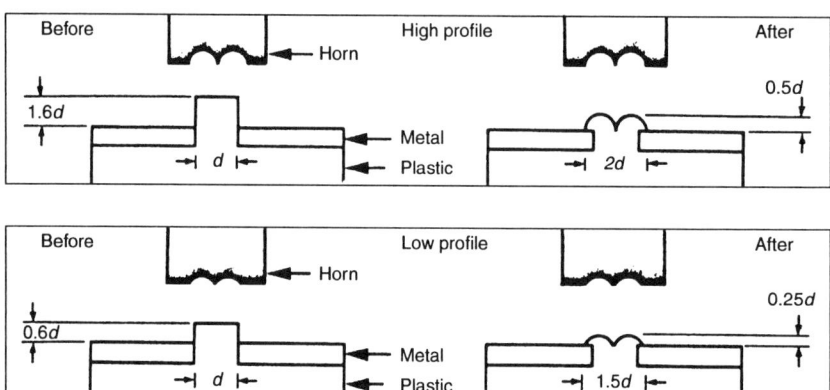

Fig. 4.12 Standard flared stake. Courtesy of Sonics and Materials Inc.

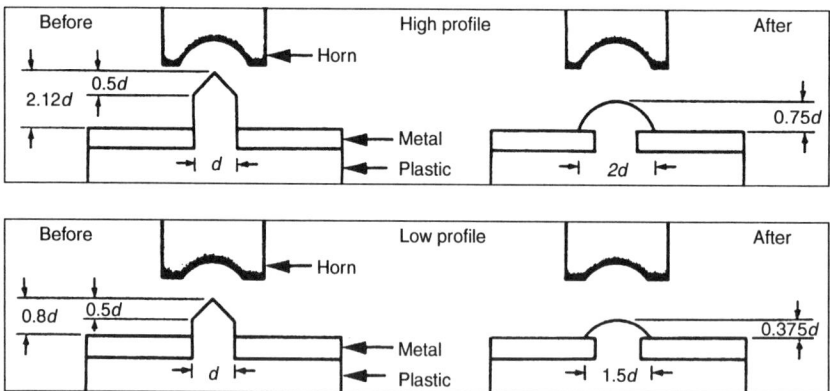

Fig. 4.13 Spherical stake. Courtesy of Sonics and Materials Inc.

Hollow Stake. Bosses with an OD in excess of 4 mm ($\frac{5}{32}$ in.) should be made hollow (Fig. 4.14). Staking a hollow boss produces a large, strong head without having to melt a large amount of material. Also, the hollow stake avoids a sink mark on the opposite side of the component and enables the parts to be reassembled with self-tapping screws should repair and disassembly be necessary.

The knurled stake (Fig. 4.15) is used in applications where appearance and strength are not critical. Because alignment is not an important consideration, the knurled stake is ideally suited for high volume production and is often recommended for use with a hand-held ultrasonic spot welder. Knurled tips are available in a variety of fine, medium, and coarse configurations.

The flush stake (Fig. 4.16) is used for applications requiring a flush surface. The flush stake requires that the retained piece have sufficient thickness for a chamfer or counterbore to be designed into the piece.

Spot Welding. Using an ultrasonic spot welder (Fig. 4.17) and standard replaceable tips, large thermoplastic parts and those with hard-to-reach joining surfaces can easily be welded together. Vibrating ultrasonically, the pilot of the tip penetrates the top sheet and enters the bottom sheet to a depth half the top sheet thickness. The displaced molten plastic is shaped by a radial cavity in the tip to form an annular formation around the weld. Simultaneously, the molten plastic displaced from the second sheet flows into the preheated area and forms a permanent molecular bond.

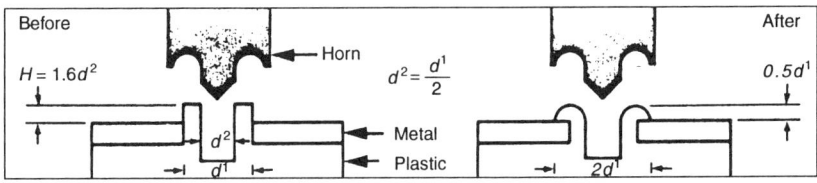

Fig. 4.14 Hollow stake. Courtesy of Sonics and Materials Inc.

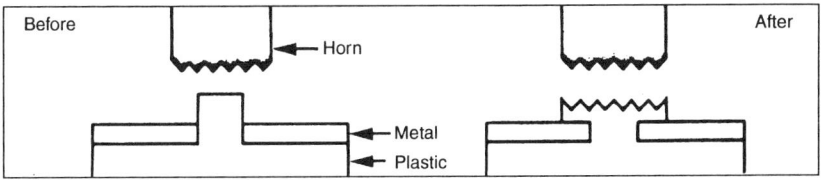

Fig. 4.15 Knurled stake. Courtesy of Sonics and Materials Inc.

62 / Decoration and Assembly of Plastics

Staking/Welding Tips. Standard threaded tips available for staking and spot welding can be obtained from ultrasonic equipment manufacturers. Special carbide-faced wear-resistant tips are available for standard horns. Horns, which cannot accept replaceable tips, can readily be carbide coated. Most frequently, bosses are ultrasonically staked one at a time using a standard horn and replaceable tip. It is possible, however, to stake several bosses simultaneously using a larger horn with multiple tips. Multielement horns can be designed to satisfy applications, where component geometry precludes the use of standard horns. Horns with up to six tips have been used successfully in multiple staking applications.

Joint Design

Perhaps the most critical facet of part design for ultrasonic welding is joint design (the configuration of the two mating surfaces). Joint configurations should be considered when the parts to be welded are still in the design stage and incorporated into the molded parts. There are a variety of joint designs; each design has specific features and advantages. Selection is determined by factors, such as type of plastic, part geometry, weld requirements, machining and molding capabilities, and cosmetic appearance.

The butt joint with an energy director is the most common joint design used in ultrasonic welding and the easiest to mold into a part. The main feature of

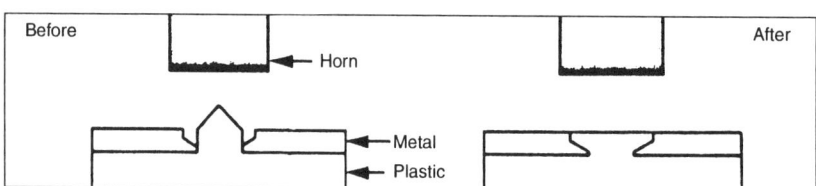

Fig. 4.16 Flush stake. Courtesy of Sonics and Materials Inc.

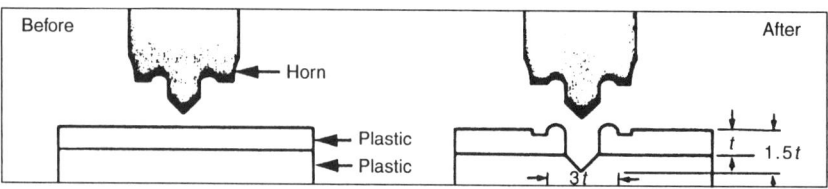

Fig. 4.17 Spot welding. Courtesy of Sonics and Materials Inc.

this joint is a small 90 or 60° triangular-shaped ridge molded into one of the mating surfaces. This energy director limits initial contact to a very small area and focuses the ultrasonic energy at the apex of the triangle. During the welding cycle, the concentrated ultrasonic energy causes the ridge to melt and the plastic to flow throughout the joint area, bonding the parts together. For easy-to-weld resins and amorphous polymers, such as acrylonitrile butadiene styrene (ABS), styrene acrylonitrile (SAN), acrylic, and polystyrene, the size of the energy director is dependent on the area to be joined. Practical considerations suggest a minimum height between 0.2 and 0.6 mm (0.008 and 0.025 in.).

Crystalline polymers, such as nylon, thermoplastic polyesters, acetal, polyethylene, polypropylene, and polyphenylene sulfide, as well as high melt temperature amorphous resins, such as polycarbonate and polysulfones, are more difficult to weld. For these resins, energy directors with a minimum height between 0.4 and 0.5 mm (0.015 and 0.020 in.) with a 60° included angle are generally recommended.

A 90° included angle energy director height should be at least 10% of the joint width, and the width of the energy director should be at least 20% of the joint width (Fig. 4.18). With thick-walled joints, two or more energy directors should be used, and the sum of the heights should equal 10% of the joint width.

To achieve hermetic seals when welding polycarbonate components, it is recommended that a 60° included angle energy director be designed into the part. The energy director width should be 25 to 30% of the wall thickness. Figure 4.19 shows a butt joint with a 60° included angle energy director.

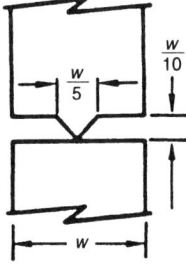

Fig. 4.18 Butt joint with a 90° included angle energy director. Courtesy of Sonics and Materials Inc.

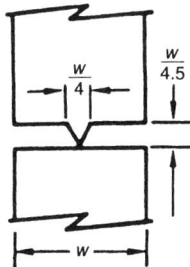

Fig. 4.19 Butt joint with a 60° included angle energy director. Courtesy of Sonics and Materials Inc.

64 / Decoration and Assembly of Plastics

Figure 4.20 shows how the assembled parts should look with the flow of the molten material from the energy director throughout the joint area.

When components are made of identical thermoplastics, the energy director can be designed into either half of the assembly. However, for maximum strength when designing energy directors into assemblies consisting of a part made of copolymers or terpolymers, such as ABS, and another part made of a homopolymer, such as acrylic, the energy director should always be incorporated into the half of the homopolymer assembly.

The step joint with energy director is shown in Fig. 4.21. This joint molds readily and provides a strong, well-aligned joint with a minimum of effort. This joint is usually stronger than a butt joint, because material flows into the vertical clearance. The step joint provides good strength in shear as well as tension and is often recommended where good cosmetic appearance is required. When working with crystalline materials, a 60° included angle energy director should be used instead of the 90° included angle energy director. Fig. 4.22 shows variations of the basic step joint design.

The tongue and groove joint with energy director (Fig. 4.23) is used primarily for scan welding, self-location of parts, and prevention of flash both internally and externally. It provides the greatest bond strength of the three joints discussed so far.

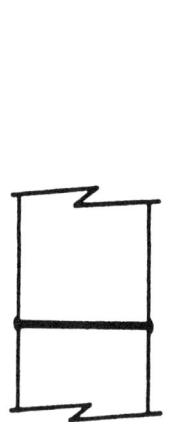

Fig. 4.20 Flow of molten material throughout joint area. Courtesy of Sonics and Materials Inc.

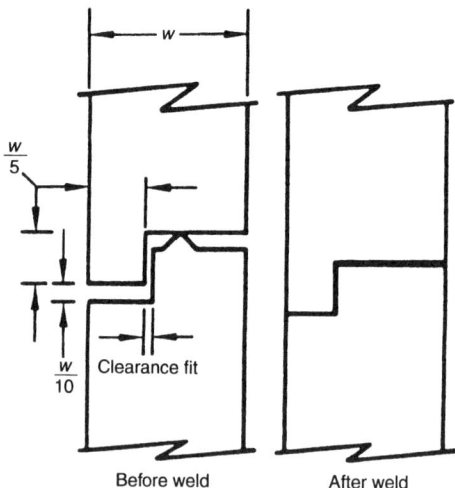

Fig. 4.21 Step joint with 90° energy director. Courtesy of Sonics and Materials Inc.

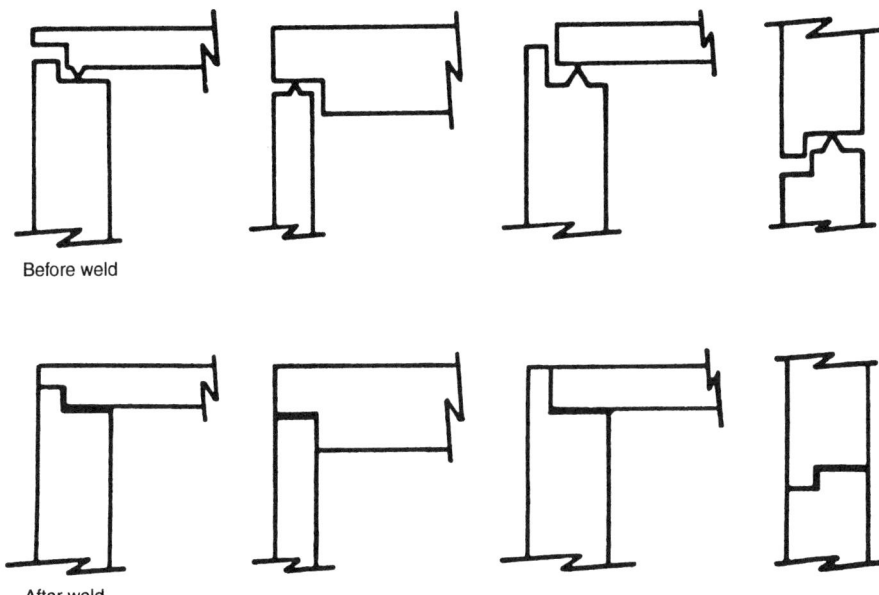

Fig. 4.22 Step joint variations. Courtesy of Sonics and Materials Inc.

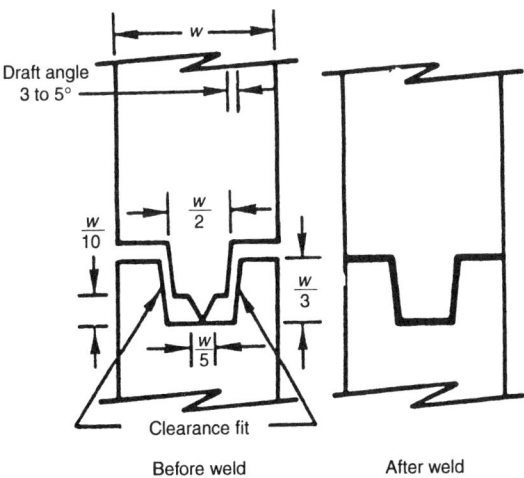

Fig. 4.23 Tongue and groove joint with energy director. Courtesy of Sonics and Materials Inc.

66 / Decoration and Assembly of Plastics

The shear joint or interference joint (Fig. 4.24) is generally recommended for high-strength hermetic seals on parts with square corners or rectangular designs, especially with crystalline resins. Initial contact is limited to a small area, which is usually a recess or step in either of the parts. The contacting surfaces melt first, then as the parts telescope together, they continue to melt along the vertical walls. The smearing action of the two melt surfaces eliminates leaks and voids, making this the best joint for strong hermetic seals.

Several important aspects of the shear joint should be considered: the top part should be as shallow as possible, the outer walls should be well supported by a holding fixture, the design should allow for a clearance fit, and a lead-in should be incorporated (Table 4.2). The shear joint requires weld times in the range of 3 to 4 times that of other joint designs because larger amounts of resin are melted. In addition, a certain amount of flash is visible on the surface after welding. When flash cannot be tolerated for aesthetic or functional reasons, a well (Fig. 4.25) should be incorporated. Figure 4.26 shows variations of the basic shear joint design. Modified joints (Fig. 4.27)

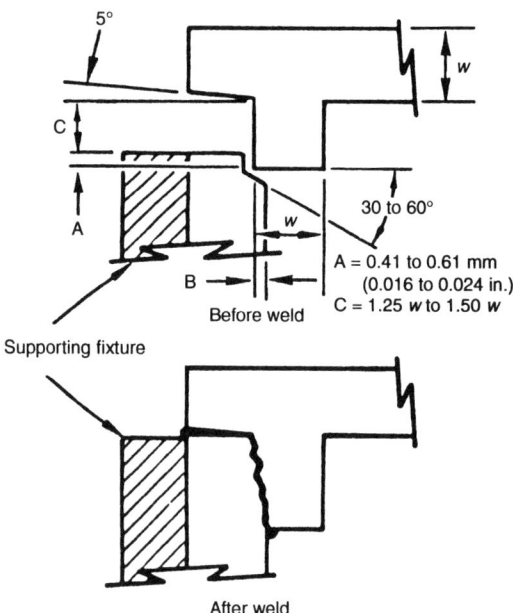

Fig. 4.24 Shear joint. Courtesy of Sonics and Materials Inc.

Table 4.2 Important considerations of the shear joint

Maximum part dimension, mm (in.)	Interference, mm (in.)
<19 (<0.75)	0.2–0.3 (0.008–0.012)
19–38 (0.75–1.50)	0.3–0.4 (0.012–0.016)
>38 (>1.50)	0.4–0.5 (0.016–0.020)

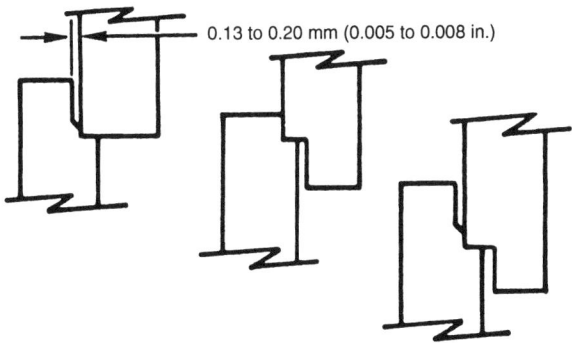

Fig. 4.25 Shear joints with flash wells. Courtesy of Sonics and Materials Inc.

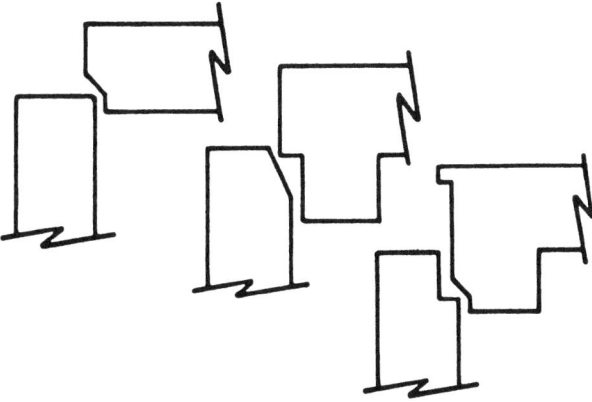

Fig. 4.26 Shear joint variation. Courtesy of Sonics and Materials Inc.

should be considered for large parts or for parts where the top piece is deep and flexible.

The scarf joint (Fig. 4.28) is generally recommended for high-strength hermetic seals on parts with circular or oval designs, especially with crystalline resins. The scarf joint requires that the angles of the two parts be between 30 and 60° and within a tolerance of 1½°. If the wall thickness is 0.63 mm (0.025 in.) or less, an angle of 60° should be used. If the wall thickness is 1.52 mm (0.060 in.) or more, an angle of 30° should be used. Intermediate angles are recommended for wall thickness between 0.063 and 1.52 mm (0.025 and 0.060 in.). A minimum wall thickness of 0.76 mm (0.030 in.) at the outer edge of the scarf is recommended to prevent "blowout," or melting clear through the wall, during welding.

The scarf joint is not commonly used due to the difficulties encountered in maintaining component concentricity and dimensional tolerances. However, this joint is highly recommended when limited wall thickness preclude

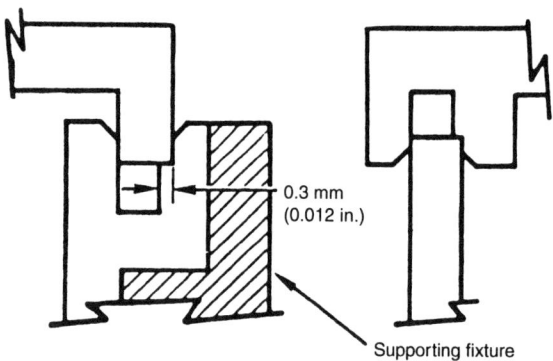

Fig. 4.27 Shear joint variations for large parts. Courtesy of Sonics and Materials Inc.

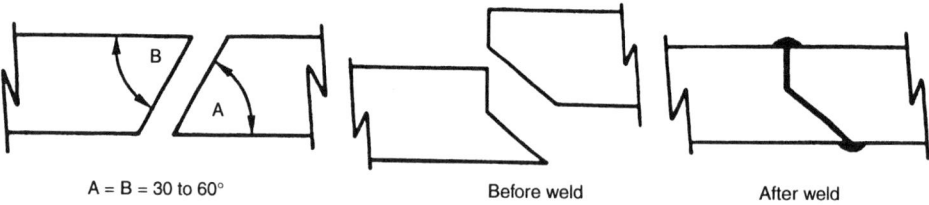

Fig. 4.28 Scarf joint. Courtesy of Sonics and Materials Inc.

the use of a shear or modified shear joint. A modified scarf joint is shown in Fig. 4.29.

As shown in Fig. 4.30, a flash well can be incorporated in the scarf joint to contain the excess molten material generated when the parts are welded. The length of the well should be at least equal to the cross-sectional thickness of the part being welded.

Ultrasonic Installation of Inserts in Thermoplastic Components

In ultrasonic insertion, a metal insert is placed in a cored or drilled hole that is slightly smaller than the insert. This hole provides a certain degree of interference and also serves to guide the insert into place. The vibrating

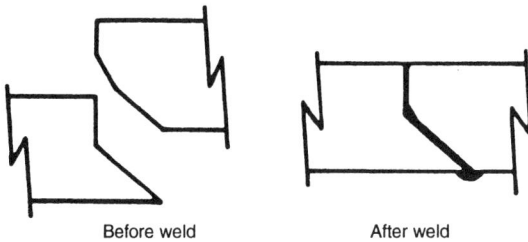

Fig. 4.29 Modified scarf joint. Courtesy of Sonics and Materials Inc.

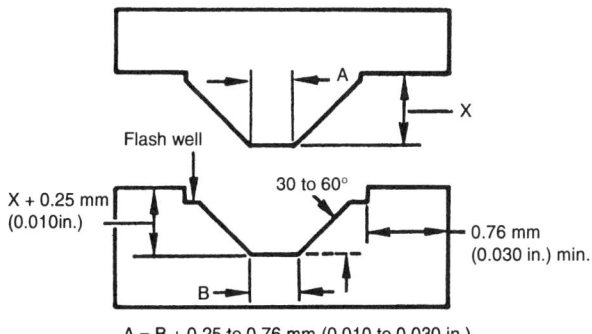

A = B + 0.25 to 0.76 mm (0.010 to 0.030 in.)

Fig. 4.30 Scarf joint with flash well. Courtesy of Sonics and Materials Inc.

ultrasonic horn contacts the insert, and the ultrasonic vibrations travel through the insert to the interface of the metal and plastic. Heat, generated by the insert vibrating against the plastic, causes the plastic to melt, and as the horn advances, the insert is embedded in the component. The molten plastic flows into the serrations, flutes, or undercuts designed into the sides of the insert. When the vibrations terminate, the plastic resolidifies, and the insert is securely encapsulated in place. In ultrasonic insertion, a slow horn approach is preferable to pressing the insert, thus allowing the horn to develop a homogeneous melt phase.

Ultrasonic insertion provides the high performance strength values of a molded-in insert, while retaining all the advantages of postmolding installation. Inserts can be ultrasonically installed in most thermoplastics. Some advantages of ultrasonic inserting include rapid installation, minimal residual stresses in the component following insertion, elimination of potential mold damage, reduced mold fabrication costs, and increased productivity as a result of reduced molding cycle times. In some applications, multiple inserts can be embedded simultaneously with special horns, increasing productivity and further reducing assembly and manufacturing costs.

Ultrasonic insertion is not restricted to standard threaded inserts. Inserts that can be installed ultrasonically include a variety of bushings, terminals, ferrules, hubs, pivots, retainers, feed-through fittings, fasteners, hinge plates, binding posts, handle-locating pins, and decorative attachments.

Typically, the plastic component is fixtured, and the insert is driven in place by the horn (Fig. 4.31). However, in some situations, the part configuration

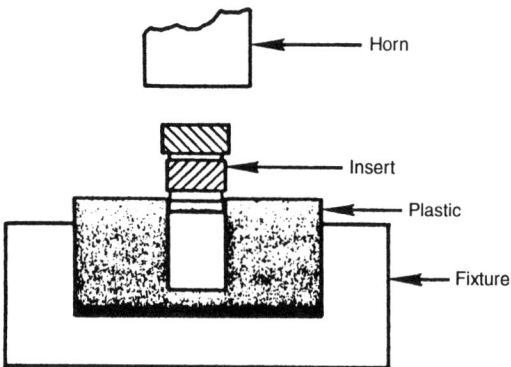

Fig. 4.31 Typical ultrasonic insertion. Courtesy of Sonics and Materials Inc.

might prohibit insert contact by the horn, and the horn is made to contact the plastic component instead of the insert (Fig. 4.32).

Functional characteristics or requirements of an application usually determine the insert design, including the side hold configuration. In all situations, a sufficient volume of plastic must be displaced to fill the undercuts, flutes, knurls, threads, and/or contoured areas of the insert. Care should be exercised in selecting the proper insert. Inserts are designed for maximum pull-out strengths, torque retention, or a combination of both. Inserts with horizontal protrusion, grooves, or indents are usually recommended for high pull-out strength requirements, while inserts with vertical grooves, or knurls, are usually recommended for high torque retention. With regard to the hole configuration or insert selection, the recommendations provided by the insert manufacturer should always be observed.

Because the horn contacts the metallic insert, it is subjected to some wear. As a result, horns used for insertion are usually made of hardened steel or titanium. For low volume applications, titanium horns with replaceable tips can be utilized. Ideally the diameter of the horn should be twice the diameter of the insert. To prevent a "jack-out" condition, the top of the seated insert should be flush or slightly above the surface of the part. Rigid fixturing should be placed directly under the insert. In most instances, it is necessary to initiate ultrasonic vibrations prior to horn contact with the insert. To maintain an accurate depth of insertion, the total distance the horn travels

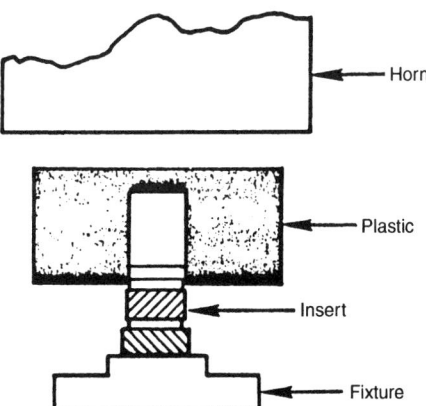

Fig. 4.32 Ultrasonic insertion that has been modified to accommodate part configuration. Courtesy of Sonics and Materials Inc.

should be limited either mechanically by a positive stop, electrically by a lower-limit switch, or both. *Caution:* When inserting, weld time should not exceed $1\frac{1}{2}$ s. Table 4.3 is a basic ultrasonic insertion troubleshooting guide.

Table 4.3 Ultrasonic insertion troubleshooting guide

Problem	Solution
Insufficient pull-out or torque strength	Decrease weld time.
	Decrease hold time.
	Decrease amplitude.
	Increase pressure.
	Increase down speed.
	Insert is too large or hole is too small.
	Improper fixturing.
	Power required exceeds capability of power supply.
Damage to insert	Decrease weld time.
	Decrease amplitude (change booster).
	Increase pressure.
	Increase down speed.
Plastic cracks	Ensure that ultrasonics is on.
	Decrease pressure.
	Walls surrounding hole are too thin.
	Increase weld time.
	Decrease amplitude (change booster).
	Decrease down speed.
	Enlarge hole diameter.
Partial insertion	Increase pressure.
	Decrease down speed.
	Decrease amplitude (change booster).
	Increase weld time.
	Increase hole depth.
	Adjust positive stop.
	Check fixturing.
	Horn is at the end of its stroke.
Inserting time is excessive	Decrease weld time.
	Decrease hold time.
	Decrease amplitude (change booster).
	Increase pressure.
	Increase down speed.
	Insert is too large or hole is too small.
	Improper fixturing.
	Power required exceeds capability of power supply.
System overloads	Decrease pressure.
	Decrease down speed.
	Decrease amplitude (change booster).
	Tune power supply.
	Check for loose studs.
	Check coupling between horn and booster.
	Power required exceeds capability of power supply.
Insert does not remain inserted	Increase hold time.
Plastic fills the threaded bore of the insert	Increase hole depth.
	Insert is too large or hole is too small.
	Insert is too long.

(continued)

Table 4.3 (continued)

Problem	Solution
Horn wears prematurely	Use hardened steel or carbide faced horn.
	Decrease amplitude (change booster).
	Insert is too large or hole is too small.
	Plastic is too abrasive.
Application is noisy	Start the ultrasonics just prior to the horn contacting the insert.
	Decrease amplitude (change booster).
	Increase pressure.
	Increase down speed.
	If possible, contact plastic rather than insert.
	Use sound enclosure or hearing protectors.
Plastic flows over the top of the insert	Adjust positive stop to limit depth of insertion.
	Decrease weld time.
	Insert is too large or hole is too small.
Horn heats up	Decrease amplitude (change booster).
	Air cool the horn.
	If possible, contact plastic rather than insert.
	Check coupling between horn, booster, and converter.

Source: Sonics and Materials Inc.

Ultrasonic Assembly System

In an ultrasonic assembly system, there are three elements in the mechanical vibrating system (Fig. 4.33):

- *The converter* produces vibrations by converting high frequency electrical energy into high frequency mechanical energy.
- *The booster horn* increases or decreases the amplitude of vibration.
- *The horn* transmits the vibrations into the plastic assembly.

These elements are half-wave sections, which vibrate about the nodal plane (Fig. 4.34).

Converter

The most efficient type of converter or transducer available is the piezoelectric type (Fig. 4.35). The converter, a sandwich type, uses a piezoelectric material called lead zirconate titanate (PZT). From a power supply, an alternating high-frequency voltage is applied to the PZT, which causes the converter to expand and contract as the polarity of the voltage changes from positive to negative. Thus, the electrical energy changes to mechanical motion. Typically, the amplitude of vibration is about 0.7 mil (0.7 μin.). To avoid energy losses, the converter is mounted at its nodal plane where there is minimal motion.

Booster

In many plastic assembly applications, it is necessary to increase or decrease the amplitude of the horn. To do this, a booster horn, a half-wave resonant section, is coupled between the converter and horn (Fig. 4.36).

The magnification ratio (the ratio of input amplitude to output amplitude) of a booster horn is determined by the cross-sectional area and shape. A horn having a stepped cylindrical shape has a magnification ratio of $(D_A/D_B)^2 = 4$. Therefore, the output amplitude at B will be (0.0007 in.)(4) = 0.0028 in. (Fig. 4.37). Lower amplitudes are obtained if the booster is reversed (Fig. 4.38). A rough approximation of amplitude magnification is

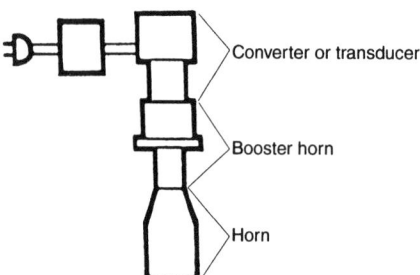

Fig. 4.33 Elements in mechanical vibrating system of an ultrasonic assembly system. Courtesy of Sonics and Materials Inc.

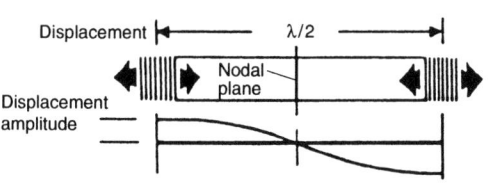

Fig. 4.34 Half-wave length sections of an ultrasonic assembly system. Courtesy of Sonics and Materials Inc.

Fig. 4.35 Typical piezoelectric converter. Courtesy of Sonics and Materials Inc.

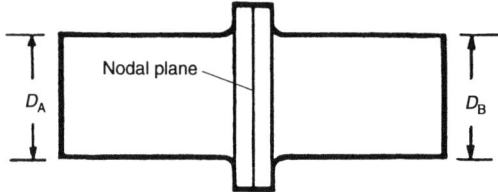

Fig. 4.36 Example of a booster horn coupling. Courtesy of Sonics and Materials Inc.

given by the ratio of the mass on the input side of the nodal plane to the mass on the output side of the nodal plane. In this situation, where the output diameter is larger than the input diameter, the amplitude at B will be (0.0007 in.)(¼) = 0.00018 in. A booster that is symmetrical about its nodal plane has a magnification ratio of one to one (Fig. 4.36).

The importance of a booster horn (amplitude) is shown in Fig. 4.39, which depicts the relationship of force applied to the horn versus power delivered from the horn. Figure 4.39 is representative of two vibrating systems with different vibration amplitudes. In a high amplitude system, as applied force is increased, power increases until the stall point (S_1), which is the capacity of the electrical power supply, has been reached. If a power supply having a capacity of 350 W is used, the power drops off at S_1, and the system stalls. However, if a power supply of 700 to 1000 W is used, higher forces can be applied until the power rises to S_2 or S_3 before the stall point is reached. In a low amplitude system, the stall points are reached at much higher applied forces.

A high amplitude system is generally used for welding delicate parts that cannot withstand high applied forces or for staking plastics where melting

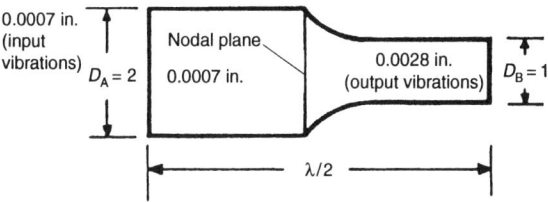

Fig. 4.37 Booster horn with stepped cylindrical shape

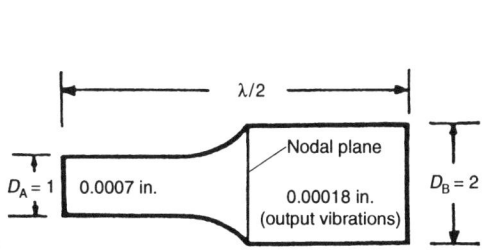

Fig. 4.38 Reversed booster horn with lower amplitudes. Courtesy of Sonics and Materials Inc.

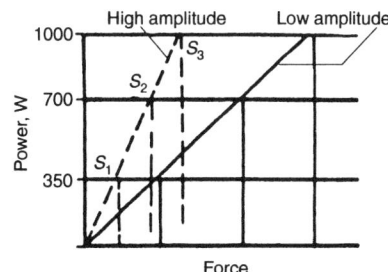

Fig. 4.39 Amplitude showing relationship of force versus power

the plastic is required. A low amplitude system is used for inserting metal into plastic. The power output (P) of a horn is given by:

$$P = \tfrac{1}{2}(FV) \cos \theta$$

where F is force, V is the velocity amplitude of the horn, and θ is the phase angle between F and V.

Velocity is directly proportional to amplitude. For simplicity, assuming perfect coupling between the horn and plastic parts (i.e., $\theta = 0$), power is proportional to force and amplitude:

$$\text{Power} = \text{Force} \times \text{Amplitude}$$

Power is sometimes given in watts (W) or in inch pounds per second (in. · lb/s). For reference purposes, 8.85 in. · lb/s = 1 W.

Horn

The horn, which comes in contact with the plastic, is designed to fit the part to be assembled. It is also designed to vibrate at its half-wave resonant frequency (20 kHz for most machinery), which is the same as the converter and booster horn. The magnification ratio of the horn, although considered in its design, is for the most part controlled by the booster horn. The half-wave length of a simple cylindrical horn is given:

$$\lambda/2 = C/2F$$

where λ is the wavelength, C is the speed of sound in the horn material, and F is the frequency of operation. For reference purposes, $\lambda/2_{Al} = 5.05$ in. and $\lambda/2_{Ti} = 4.99$ in.

The speed of sound in the horn material (C) is:

$$C = \sqrt{E/\rho}$$

where E is the modulus of elasticity and ρ is density. Horns other than a solid cylindrical design do not lend themselves to this formula. Using it as a first approximation, however, and using past empirical data, horns can be simply tuned to operate at the half-wave length resonant frequency.

In considering what material should be used for a horn, two parameters are important: low acoustic impedance and high fatigue strength. The acoustic impedance of the horn material determines the power loss in the horn. Lead and copper are examples of metals having high impedance, and titanium and aluminum have low impedance. Impedance is given by:

$$Z = \rho C$$

$$Z_1 = \rho C A$$

where Z is the characteristic acoustic impedance, Z_1 is specific acoustic impedance, ρ is density, C is the speed of sound, and A is the cross-sectional area.

The fatigue strength of the material determines the maximum velocity or amplitude that can be obtained without failure from fatigue cracking the metal. This maximum velocity is given by:

$$V_{MAX} = F/\rho C$$

where V is horn velocity amplitude, F is the fatigue strength, and C is the speed of sound.

Another consideration for horn material is wear resistance or hardness. In assembling plastics, the horn is subjected to the abrasive action of the plastic. Aluminum, considered to be a good horn material as far as fatigue strength and impedance characteristics are concerned, has very poor wear characteristics. Aluminum horns tend to leave a black oxide mark on a plastic assembly. If an aluminum horn is used, the horn should be chrome plated or hard coated. Aluminum horns should not be stressed over 2.5 mils (0.0025 in.) at 20 kHz. Titanium should be used for applications that require amplitude over 2.5 mils (0.0025 in.). For diagnostic purposes, a cracked horn will have a high reading on the horn-tuning instrument.

Horn Velocity. The calculation of the horn velocity gives an appreciation of the energy involved. For example, a 20 kHz welder with a tip displacement of 1.5 mm (0.005 in.) has a horn velocity defined as the frequency times displacement, or 30,000 mm/sec (1,000 ft/sec). If such a horn had 379.2 to 413.7 kPa (55 to 60 psi) air pressure forcing it down at this velocity, it would equate to about 4086 kg (9000 lb) being dropped 25.4 mm (1 in.). In order to effectively and efficiently deliver the vibration to the horn, a few requirements must be considered:

- The booster and horn should be equivalent to one wavelength of the welding operating condition. The horn length should be half the wavelength.
- The booster and horn should be made using high attenuating metals, such as titanium or aluminum.
- The booster/horn assembly must be tuned.

Horn tuning is often conducted at the factory where the horn is manufactured, because it requires special equipment. An oversimplified (but effective) example of the tuning process is the tuning of a guitar. The strings have to be a certain material and a certain length, but they must also have a certain tension. Some musicians, for example, tune their instruments by ear while others utilize electronic tuning systems. The electronic tuning system for musical instruments is not unlike that for tuning ultrasonic horns. Using the same analogy, as the length or tension of a guitar string can not be altered without affecting the tuning of the guitar, so an ultrasonic horn cannot be significantly altered (via machining or wear) without affecting its function. Usually an out-of-tune horn results in a loud squeal, an inability to weld, or both.

Dielectric Sealing

Dielectric sealing is also referred to as radio frequency (RF) sealing. Special equipment is required, and only a few plastic materials, such as flexible vinyl, can take advantage of this process. The equipment can be described as an industrial microwave generator. Weld plates are connected to a circuit, which includes a cavitron tube or RF generator. What actually occurs is an alternation in the charge or polarity of the weld plate. Two weld plates are opposite in charge, and the positively charged plate is changed to a negative charge and vise versa at a rate that is the function of the frequency of the RF generator (Fig. 4.40). As the charges are switched, the polar plastic material will tend to have its molecules align with the changing polarity. As the frequency of the change in polarity increases, the moving molecules rub against each other, and the resulting friction generates sufficient thermal energy to soften the plastic. At the appropriate moment as defined by the softening temperature of the material, the RF energy ceases, and the two weld bars close on the plastic, squeezing them together, creating the bond, and simultaneously acting as a weld chill cooling the plastic at the bond area.

Material Compatibility. The dielectric sealer requires the plastic(s) to be bonded to be intrinsically polar. The most commonly used polar plastic is plasticized polyvinyl chloride (PVC), which is commonly a film or sheet.

Applications. Dielectric sealing has found a niche in several unique applications including vinyl covered fiberboard used to produce notebooks, inflatable beach and pool toys, waterbed mattresses, and similar bladder applications.

Advantages and Disadvantages. As with other plastic welding processes, dielectric sealing does not require any third component materials, such as adhesives, which allows for simpler manufacturing and excellent recyclability. The main disadvantages include that only polar materials can be dielectrically sealed, thus limiting the material choices. Also, the dielectric sealing equipment is unique and requires special investment.

Joint Area Design. There are no real special requirements for joint designs. Flat sheet or film is adequate for bonding.

Induction Inserting

One of the most common applications of plastic reflow is the installation of threaded inserts. Most thermoplastics are too soft to sufficiently hold a

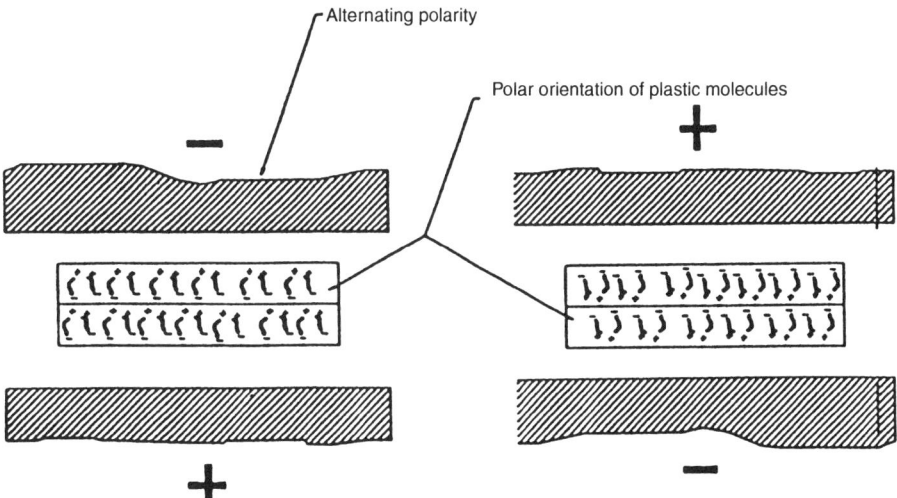

Fig. 4.40 Dielectric sealing

thread, so brass or steel threaded inserts are added. Postmolded installation is more cost effective than molding in place, and induction is a proven way to preheat the inserts prior to installation. Other applications for bonding plastic to metal include the insertion of cutlery into handles and ball-type grips onto shift levers.

The insert is preheated and then pressed into a hole in the plastic part by positioning a coil where the insert is to be pressed. The insert is then held in the coil during the heating cycle. When the correct temperature is achieved, the insert is pressed into the plastic (Fig. 4.41, 4.42). A narrow zone of plastic then melts and flows into the knurls of the inserts. The plastic cools

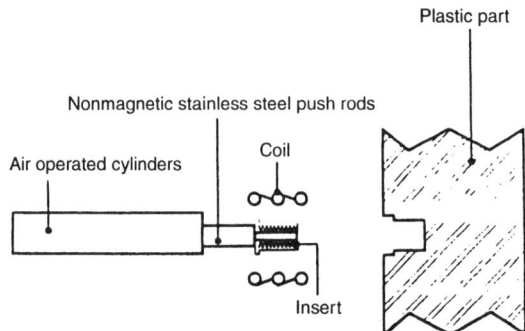

Fig. 4.41 The insert is held in the coil during the heating cycle. Courtesy of Ameritherm Inc.

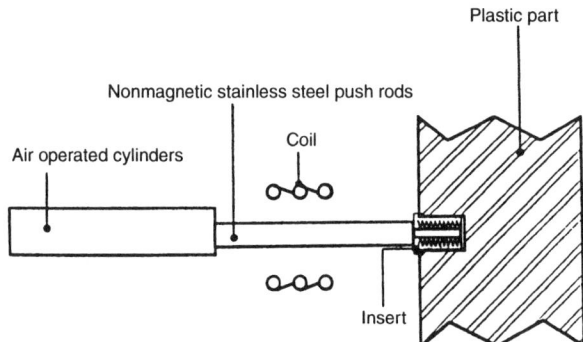

Fig. 4.42 The insert is pressed into the plastic at the correct temperature. Courtesy of Ameritherm Inc.

and resolidifies, providing a complete assembly with much better mechanical properties than inserts placed using other techniques.

Inserting Metal into Plastic

Insert material is usually brass or steel, and each of these has advantages and disadvantages. Brass is nonmagnetic, and it will not corrode as easily as steel. However, brass is a softer material than steel is; it will anneal at lower temperatures than steel. Brass starts to anneal at 232 °C (450 °F), while steel starts at 649 °C (1200 °F). Some glass-filled plastics require inserts to be 371 °C (700 °F) for correct installation; therefore, the brass inserts must be heated and inserted fast to prevent thread annealing. Also, brass takes more power to heat with induction than steel.

Steel inserts are harder than brass, and that means there is no annealing concern at the insertion temperatures of most plastics. A steel insert can be heated faster and easier with induction. The consistency of the coating or plated finish of steel inserts, particularly in the threads, can be difficult to achieve. Steel inserts are typically magnetic and oxidize faster than brass. Selection of the insert material is dependent on the specific application requirements. However, brass tends to be the material of choice due to the coating and plating issue.

Insertion Installation. There are some considerations when using induction for installing inserts in plastic. The insert must have the proper knurl and fin design to achieve the desired rotational torque and tensile strength (Fig. 4.43). Most inserts designed for molded-in, expansion, ultrasonic, or self-threading insertion can be installed with the induction heating process. However, due to the ability of the inductively heated insert to easily reflow plastic, the knurls and fin on the insert can be made deeper for greater

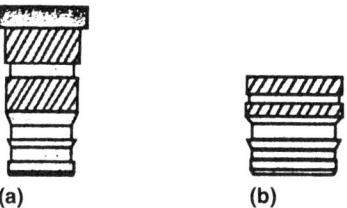

Fig. 4.43 Knurl and fin design for (a) flanged and (b) flush inserts. Courtesy of Ameritherm Inc.

holding strength. The material and mass of the insert with the available heating time determine the power required to reach the desired temperature.

The installation temperature of the insert is important. Each insert must be heated to the same temperature within the same time to achieve a consistent process. The critical parameters for a consistent induction insertion process include (Table 4.4):

- The power on cycle in the induction power supply
- The heat on time
- The tune frequency
- The power ramp up and ramp down
- The positioning of the insert in the coil
- The insertion pressure
- The thermoplastic material

Thermoplastics will flow at elevated temperatures, and the solidified polymer can be reheated to reflow around the insert. With thermosetting polymers, once the shape has been cast, they no longer melt or flow on reheating. Table 4.4 shows some examples of typical temperature, time, and power required for different inserts and plastics.

The diameter of the hole for the insert must be the correct size to allow the plastic to flow around the insert. If the hole is too small, the extra plastic will have to be displaced. The displaced plastic causes what is known as "flash." Excessive heat and pressure may also cause flash. If the hole is too large, not enough plastic will flow into the fins and knurls to achieve the desired holding strength.

Table 4.4 Critical parameters for a consistent induction insertion process

Quantity	Material	Size of insert, in.	Temperature, °C (°F)	Heat time, s	Total power delivered, kW	Total power required, kW
2	Brass	0.25 diam × 0.50 long × 0.12 wall thickness	177 (350)	2	0.8	1
1	Brass	0.25 diam × 0.63 long	204 (400)	2.5	1.2	3
1	Brass	0.33 diam × 0.50 long × 0.10 wall thickness	204 (400)	0.8	0.3	1
1	Brass	0.33 diam × 0.50 long × 0.17 wall thickness	204 (400)	1	0.3	1
1	Brass	0.56 diam × 1.25 long	371 (700)	6	0.8	1
4	Steel	0.63 diam × 0.50 long × 0.12 wall thickness	260 (500)	12	0.6	1
1	Steel	0.75 diam × 0.05 long	121 (250)	2	0.7	1

Source: Ameritherm Inc.

With proper fixturing, tight center-to-center tolerances of the inserts can be held. Molded plastic parts sometimes do not have tight tolerances. Postmolded installation, using induction heating and accurate insert location tooling, will put the insert in the same location each time even if the hole dimensions in the plastic part vary.

Insertion Method. There are several methods used for inserting with induction heat. An x-y positioning table can be used in conjunction with a single position coil when multiple inserts need to be installed in a single molded part (Fig. 4.44). The position of the coil is held constant, and the x-y table moves each insert location under or above the coil. This technique provides for a flexible manufacturing tool that can be changed by software programming, rather than hardware tooling changes.

A second option for a single position coil is the use of a robotic arm (Fig. 4.45). The plastic part is held in a fixed location, and the coil and indexing mechanism can be moved to each insert location. Encapsulated heat stations are available from Ameritherm Inc. (Rochester, NY) for this process. Small encapsulated heat stations can be supplied with 1, 3, 5, and 7.5 kW power supplies with mounting brackets that align with robotic insertion tooling. Location of the coil relative to the heat station is custom designed for each application.

A multiple position coil makes it possible to install more than one insert at a time into a single plastic part. Three and four position coils have been

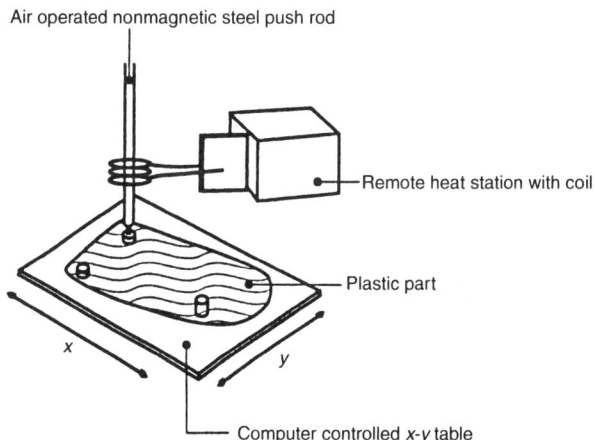

Fig. 4.44 Stationary inserter, variable x-y table. Courtesy of Ameritherm Inc.

84 / Decoration and Assembly of Plastics

used for this application. For this arrangement, the coil is normally in a fixed position, and the plastic part is moved to the coil. The inserts for each location are heated simultaneously and then pressed into the plastic (Fig. 4.46).

For all tooling arrangements, the manner in which the insert is held is more important with induction than with other insertion methods. Having metallic tools hold the insert in the coil impedes the performance of the induction power supply. The tooling in direct contact with the inserts should be nonmagnetic and have the smallest mass possible. Nonmagnetic stainless steel is often used as the rod to hold and insert a part. The insert threads should not be used for holding, which can damage the threads during insertion. Three jaw chucks or a combination of a locating rod and vacuum can be used for holding the insert. With each method, a taper or some other centering mechanism should be incorporated into the design. Gravity is often used to position the inserts on the insert tooling, and the inserts are normally pushed up into or horizontally into the plastic.

Metal to Plastic Bonding

Steel, brass, and sometimes aluminum are the materials of choice for metal inserts for plastic. The relative merits of each material and the effect on the ability to be heated and inserted into the plastic part should be

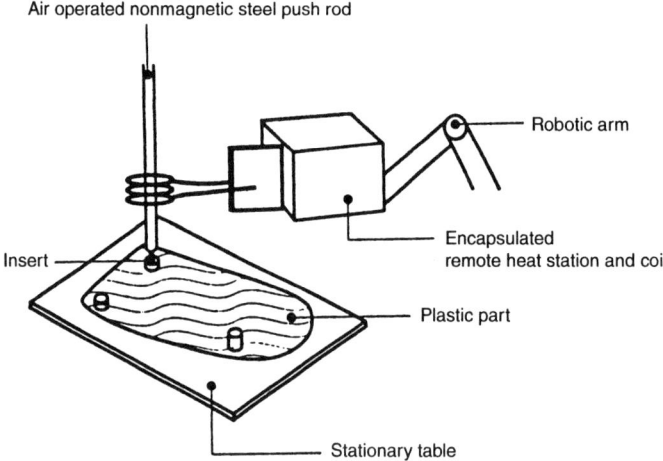

Fig. 4.45 Stationary table, variable position inserter. Courtesy of Ameritherm Inc.

considered. Material resistivity, permeability, specific heat, and thermal conductivity are the four primary properties of the material that are important when induction heating the metal insert. All these materials can be heated by induction, but the material properties dictate the characteristics of the induction process.

Table 4.5 shows the characteristics of 9.5 mm (⅜ in.) OD inserts made from aluminum, brass, and steel being heated with the same magnetic field produced by a coil with the same current and the same number of turns. The

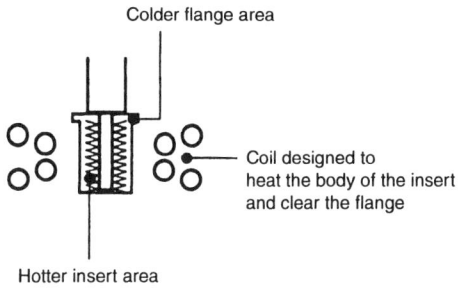

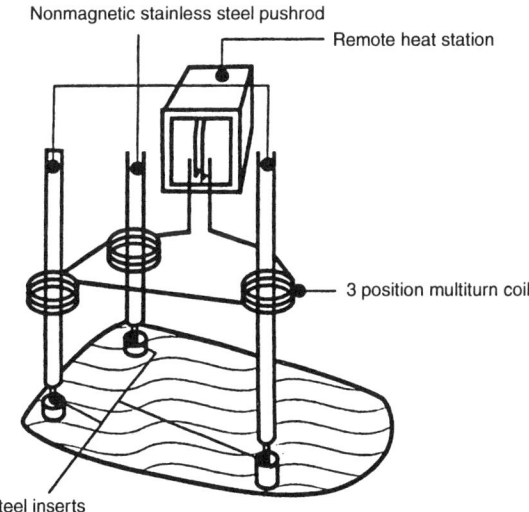

Fig. 4.46 Multiposition coil. Courtesy of Ameritherm Inc.

steel insert absorbs more power from the magnetic field; therefore, it is heated faster than the aluminum and brass inserts.

Effects on Properties. The induction process produces heat in the insert by creating an electrical current that flows around the part. This induced current flows through the resistivity of the material causing heat to be generated within the insert. The current flows toward the outside surface of the insert with most (80%) flowing in an area known as the "skin depth." The skin depth is dependent on the resistivity and permeability properties of the material. Table 4.5 shows a steel insert with a skin depth of 0.0635 mm (0.0025 in.) compared to brass (0.203 mm, or 0.008 in.).

With a steel insert, the heat is produced closer to the edge, so care must be taken not to melt the outer surface of the insert during fast heating cycles. The heating time must be long enough to allow sufficient energy to be transferred to the insert for proper bonding when it is inserted into the plastic. The 0.203 mm (0.008 in.) skin depth of the brass insert produces heat deeper in the insert but dictates that small diameter inserts must be heated with higher frequencies to maintain heating efficiently. For optimal efficiency, the diameter of the insert must be greater than four times the skin depth. The skin depth (d) is given by the fundamental equation:

$$d = 3160\sqrt{\rho/\mu f}$$

where ρ is the resistivity, μ is the permeability of the metal, and f is the operating frequency (Hz). In order to achieve efficient heating of small inserts, the operating frequency of the induction equipment must be above 50 kHz. When considering the overall size of the insert, the available insert length for an effective coil, the resulting small number of turns, and low coil

Table 4.5 Inserts with 9.5 mm (3/8 in.) OD in same magnelic field in a coil with same number of turns and same current

Property	Aluminum	Brass	Steel
Resistivity, $\mu\Omega \cdot m$ ($\mu\Omega \cdot in.$)	0.028 (1.11)	0.07 (2.76)	1.566 (29.0)
Skin depth, mm (in.)	0.127 (0.005)	0.203 (0.008)	0.0635 (0.0025)
Specific heat, cal/g °C	0.214	0.092	0.118
Thermal conductivity, W/m · K	211	395	52
Temperature, °C (°F)	177 (350)	177 (350)	177 (350)
Time to heat, s	3.2	3.0	1.1
Power absorbed by insert, W	350	484	560

Source: Ameritherm Inc.

inductance (operating frequencies, 300 to 450 kHz) are required for efficient coil/part energy transfer.

Table 4.6 shows the amount of power required to heat 9.5 mm (3/8 in.) diameter inserts to 177 °C (350 °F) in 1 s. The amount of energy required to heat the insert depends on the mass of the insert and the specific heat of the material described by the following equation:

$$Q = mC_p \Delta T$$

where Q is the quantity of heat (kW), m is mass, C_p is the specific heat, and ΔT is the rise in temperature.

The power required by the insert is a direct relationship to the time taken to heat it. If the heating time increases from 1 to 3 s, the amount of power required decreases to 205 W. Although the power required to heat the brass and aluminum inserts is less than steel, the power absorbed from the magnetic field is far less than the steel insert (Table 4.5). Thus, the steel inserts heat much more efficiently than the brass or aluminum and require much less power from the power supply. Therefore, a 1 kW induction power supply can be used to heat one aluminum, two brass, or three steel inserts at a time with a properly designed single or multiposition coil.

Material Insertion Process. Specific heat and thermal conductivity are the primary properties to consider in the material insertion process. Because the induction process produces heat in the insert toward the outside of the part, the heating pattern should be considered relative to the process time.

At the end of the heating cycle, the surface of the insert is at a relatively higher temperature than the inside diameter (ID). However, due to the thermal conductivity of these materials, the insert quickly reaches a uniform temperature throughout. Heating the 9.5 mm (3/8 in.) OD brass insert to 177 °C (350 °F) in 1 s will only take the center thread surface 60 ms to reach temperature equilibrium. The steel insert will take as long as 25 ms.

Table 4.6 Conditions required to heat 9.5 mm (3/8 in.) OD inserts to 177 °C (350 °F) in 1 s

Condition	Aluminum	Brass	Steel
Mass of insert, kg (lb)	0.0026 (0.0058)	0.0083 (0.0184)	0.0075 (0.0165)
Specific heat, cal/g · °C	0.214	0.092	0.118
Temperature rise, °C (°F)	149 (300)	149 (300)	149 (300)
Power, W	390	535	615

Source: Ameritherm Inc.

88 / Decoration and Assembly of Plastics

As time and temperature effect the internal screw threads, a short heat cycle and dwell time prior to insertion will minimize the possibility of thread annealing. Figures 4.47 and 4.48 show the temperature change through the insert over time when heated by induction using a 1 kW power supply.

After the heating cycle, during the insert location and before actual insertion, the insert loses heat by radiation and convection. Heat is also lost to the fixture holding the insert. The drop in surface temperature is almost exponential (Fig. 4.49). It is, therefore, important that the insert be pushed into the plastic as soon as possible after the end of the heating cycle. With a brass insert, it must be inserted within 5 s of the end of the heating cycle, otherwise the insert loses sufficient heat to melt and flow the plastic.

Figure 4.49 shows the surface temperature of the brass insert when it is pushed into the plastic. The temperature of the insert drops much more rapidly after insertion, because it loses heat in melting the plastic and by conduction into the plastic. A comparable curve for steel is shown in Fig. 4.50. Steel inserts can be heated very quickly, and due to the poor thermal conductivity of steel, the surface temperature of the steel is much higher than the center of the insert. Figure 4.50 shows that at the end of the heating cycle, the temperature of the surface of the insert drops rapidly as the heat

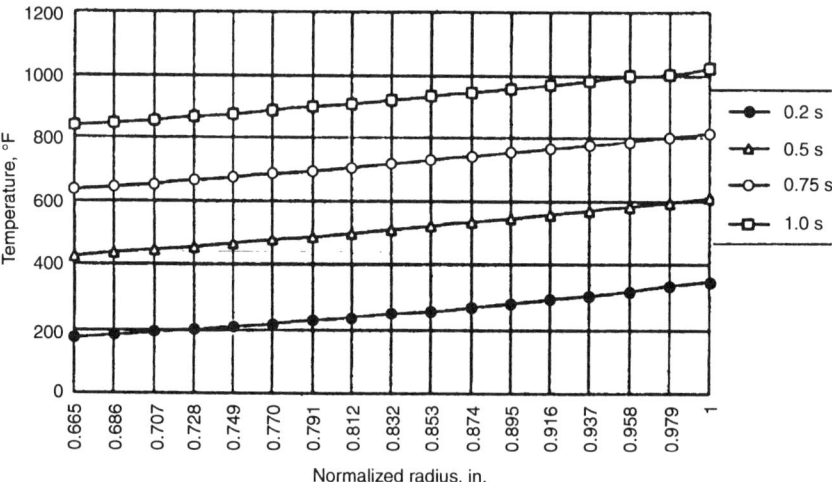

Fig. 4.47 Temperature through the cross section of a steel insert with a 9.5 mm (3/8 in.) OD. Courtesy of Ameritherm Inc.

Welding Assembly of Plastics / 89

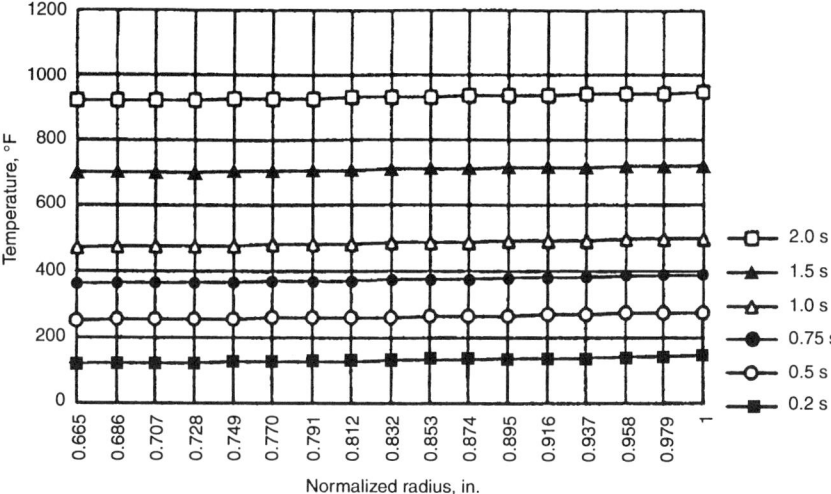

Fig. 4.48 Temperature through the cross section of a brass insert with a 9.5 mm (3/8 in.) OD. Courtesy of Ameritherm Inc.

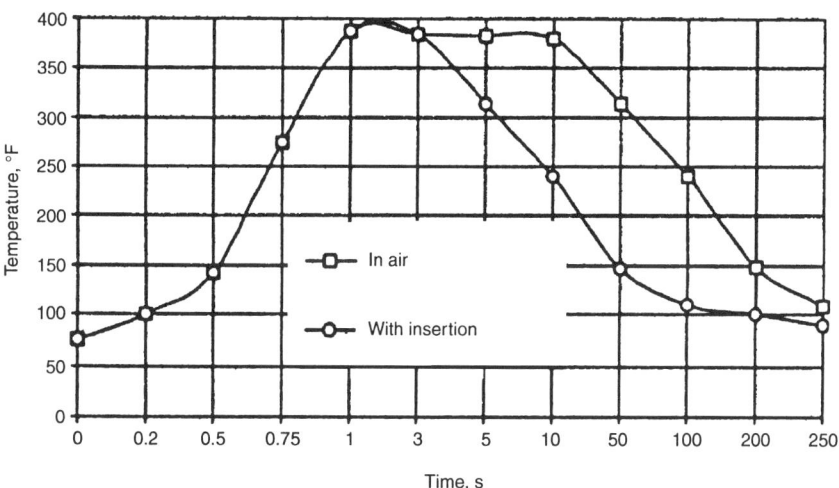

Fig. 4.49 Temperature of a 9.5 mm (3/8 in.) brass insert in air and when inserted in plastic. The insert is heated for 1 s and should be inserted within 5 s to obtain good capture of the plastic around the insert. Courtesy of Ameritherm Inc.

travels to the center of the insert until the whole insert reaches the equilibrium temperature 204 °C (400 °F).

The heat content of the metal insert just before it is pushed into the polymer should be sufficient to melt the polymer to a plastic state and produce a good flow around the insert. The interference between the insert and the guiding bore in the polymer determines the amount of polymer to be displaced for proper insertion. It is usually not sufficient to heat the insert to the melting point of the plastic. The insert must have sufficient heat content to supply the necessary heat to raise the temperature of the polymer at the interface to the melting temperature and also to overcome the latent heat of fusion of the polymer. The latent heat of a substance is the amount of heat required at the melting temperature to change the phase from solid to liquid. This temperature depends on the particular plastic used.

A brass insert at 177 °C (350 °F) could not be inserted into a plastic that melts at about 93 to 127 °C (200 to 260 °F). The insert must be heated to about 204 °C (400 °F) to get a good insertion. The insert must be heated above the melting temperature of the plastic to get a good flow around it. However, the insert must not be too hot, or the plastic will burn or boil, which can create excessive flashing.

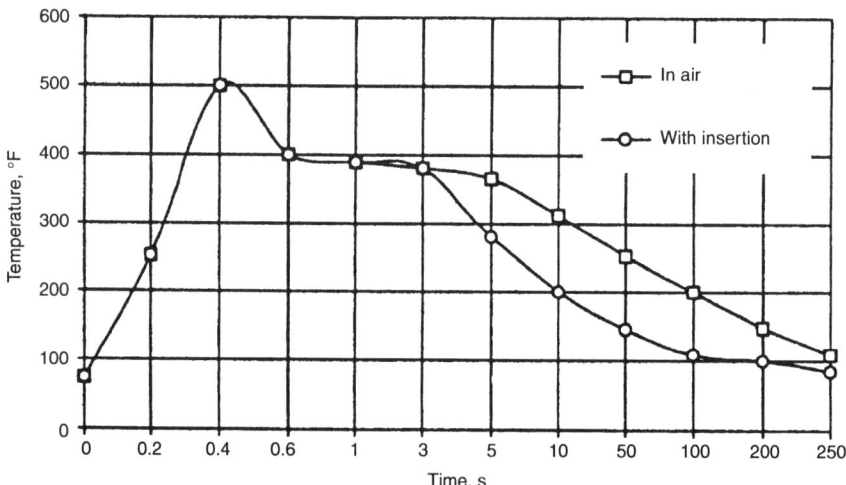

Fig. 4.50 Temperature of a 9.5 mm (3/8 in.) steel insert in air and when inserted in plastic. The insert is heated for 0.4 s and should be inserted within 3 s to obtain good capture of the plastic around the insert. Courtesy of Ameritherm Inc.

Once the insert has been pushed into the polymer, a temperature gradient exists in the polymer. The overall temperature is the hottest at the insert/polymer interface, and it drops off exponentially away from the interface. The locating fixture should be held stationary after insertion to allow the polymer to resolidify around the insert. For the 9.5 mm (3/8 in.) brass insert, a hold time of 0.5 s is usually sufficient to allow the plastic to resolidify. This allowance will also maintain the tolerance required by the insert. The hold time depends on a number of factors, such as the insert material, the insert temperature at the time of insertion, the amount of polymer displaced, and the temperature characteristics of the polymer. The melting point and viscosity of the polymer determine the required temperature of the insert for satisfactory insertion.

The process of inserting the metal insert in the plastic depends on the material characteristics of both the insert and the plastic. During the heating process, an optimal coil design is needed to transfer power efficiently and quickly to the insert. Steel, brass, or aluminum can be used for inserts, although steel heats the easiest. Heat lost by radiation and convection, as well as the heat lost by conduction to the inserting fixture, must also be addressed during the design of the process. A good capture of plastic is vital for a good pull strength and resistance to torque. Thus, an optimal insertion process depends on a properly designed coil, metal insert, and the guiding hole for the insert. The insert must also be heated to the right temperature to have a good strong product with good capture of plastic around the metal insert (Fig. 4.46).

Simultaneous Insertion of Three Steel Inserts

There are many plastic molding applications that need the assembly of two or three inserts. By simultaneously heating the inserts with induction in a multiposition coil, the inserts can be pushed into the part at the same time optimizing the part throughput through the insert machine. For example, an automotive door handle requires three steel inserts to be placed in three separate locations on the handle. The locations are on different planes, but insertion is from the same side. The steel inserts heat well and are easy to load with induction energy. A multiposition coil designed to the same hole positions as the inserts on the handle heats the three steel parts to 191 °C (375 °F) in 2 s. The coil is in a single plane, and the insert push rods are designed to correctly place each insert into the door handle on a 5 s cycle. These steel inserts have a 1/2 in. diameter flange, which must seat on the door handle but not melt into the surface. By designing the coil correctly and positioning the part in the coil, it is possible to heat the body of the insert to a higher temperature than the flange.

Inserting Metal Parts Accurately

It is often necessary to locate the metal inserts in the molded plastic part with greater accuracy than is achievable by ultrasonic insertion techniques. Errors in the precise location of the holes in some plastic moldings cause parts inserted ultrasonically to "wander" from the required location. By using induction heating to heat the insert and with accurate insert location, it is possible to correct the initial location error by placing an additional insert in the hole in the molded part. As the heated insert causes the plastic to melt and flow, a 0.762 to 0.254 mm (0.03 to 0.10 in.) correction in the location of the insert is achievable (Fig. 4.51).

For example, an air nozzle requires two inserts to be accurately positioned relative to the part face independent of the molded hole spacing. By accurately locating the insert push rods relative to the part with a loosely coupled two-position coil, the inserts could be precisely inserted into the air nozzle molding using a 1 kW power supply. Heating the two inserts in approximately 2 s with the insert and cure time taking an additional 3 s gives a total cycle time for one part of 6 s. The inserts are heated to 177 °C (350 °F), and the system operates at a frequency of 207 kHz with 275 W loaded into both inserts.

Applications

Chair Frame. The office environment of today uses plastics in office furniture. Chairs often have a metal leg assembly with a molded plastic chair body attached with steel fasteners. Metal inserts are screwed into the plastic chair body with the leg assembly bolted to it. Unfortunately, the retention and torque strength of the screw in metal inserts is not very reliable, and the assemblies over time tend to loosen.

If induction heat is used, the screw becomes an integral part of the plastic chair body (Fig. 4.52). The steel insert is heated sufficiently to melt the plastic upon insertion. The plastic reflows around the knurls, solidifies, and forms a lasting bond. Once it has solidified, the retention and torque strength is extremely high, and the process can be duplicated quickly and repeatedly.

In this application, four steel inserts are heated simultaneously with a multiturn, multiposition coil. The inserts are positioned in the coil using nonmagnetic stainless steel push rods and heated to 316 °C (600 °F) in approximately 13 s before insertion into the plastic chair body. The insert temperature of 30% glass-filled plastic is 254 °C (490 °F). The induction heated insert process has a much higher retention strength and positional accuracy than the screwed-in insert.

Silverware. The cutlery industry has many requirements for assembling plastic handles onto steel flatware. Designs for pocket knives, dinner knives,

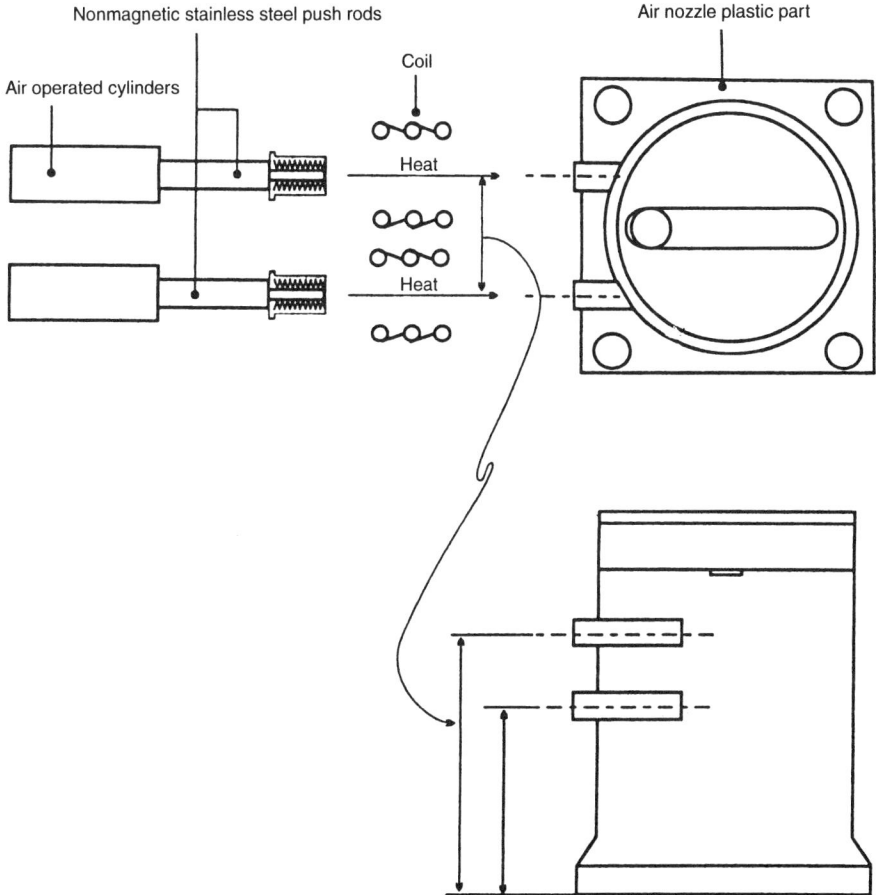

Fig. 4.51 Air nozzle with two inserts positioned relative to the part face, independent of the molded hole spacing. These dimensions must be maintained even if the molded hole is off by 0.005 in. Courtesy of Ameritherm Inc.

94 / Decoration and Assembly of Plastics

forks, and spoons all use plastic handles. Induction heat is used to heat the tang of the flatware either before or after assembly to the plastic handle.

In this application (Fig. 4.53), a plastic handle is positioned on a pancake coil, and the flatware tang aligned at the end of the handle (opposite end of the coil). Power is applied to the coil, and the tang/handle is moved to the coil center position. The tang is pushed into the handle as the ABS melts, and the entire assembly is removed from the coil as the tang fully seats. The time to complete the operation is 4 s. This application operates at 270 kHz and delivers 200 W of power raising the tang to approximately 177 °C (350 °F). This temperature allows the plastic to form correctly around the tang, but it is not hot enough for the handle to lose its molded shape. The pancake coil is used so the staking process can be directly incorporated into the assembly line.

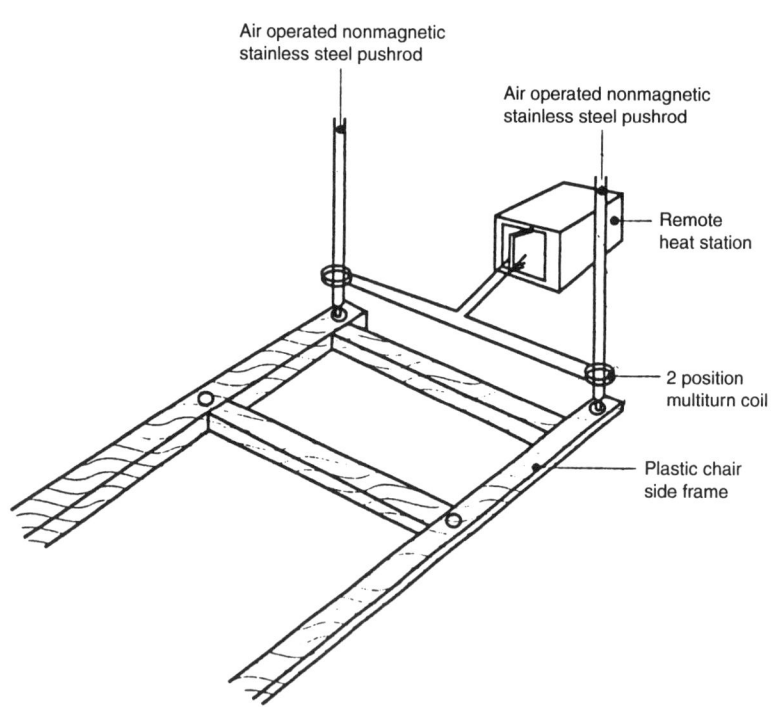

Fig. 4.52 Induction-heated insert as part of a plastic chair body. Courtesy of Ameritherm Inc.

Plastic Bearings onto Steel Shafts. With the advances in polymer applications for bearing surfaces, there are many applications needing the attachment of a bearing assembly to a steel shaft. Using induction to heat the end of the steel shaft, the shaft can be inserted into the bearing to form a robust low-friction bearing surface. Plastic bearing assemblies are often used on the end of air spring cylinders in the automotive industry to aid the opening of hoods and trunks.

In this application (Fig. 4.54), the ends of two steel shafts are heated simultaneously in a multiposition coil using a 3 kW power supply with an encapsulated remote heat station. The heat station and coil assembly move over the two steel shafts and heat the ends of the shafts to 704 °C (1300 °F) in less than 2 s before moving back from the shafts. The plastic bearing assemblies are placed on the ends of the shaft; 880 W are delivered to each shaft end; and the total cycle time takes approximately 15 s. The high temperature of the ends allows for cooling after the shafts are removed from the coil, the coil is moved back, and the plastic bearings are inserted. By keeping the steel shaft and the bearing in one vertical plane and moving the coil and encapsulated heat station for the heat cycle, a more reliable automated process is achieved.

Electrical Connectors. Many two part electrical connectors are held together using a mechanical threaded screw to make sure they do not separate. A 6.35 mm (0.25 in.) OD brass insert is often used to provide the robust thread in the plastic connector body. This insert can be precisely heated using induction (Fig. 4.55) to give a very repeatable process. The 6.35 mm (0.25 in.) OD, 15.9 mm (0.625 in.) long brass insert is heated to 316 °C (600 °F)

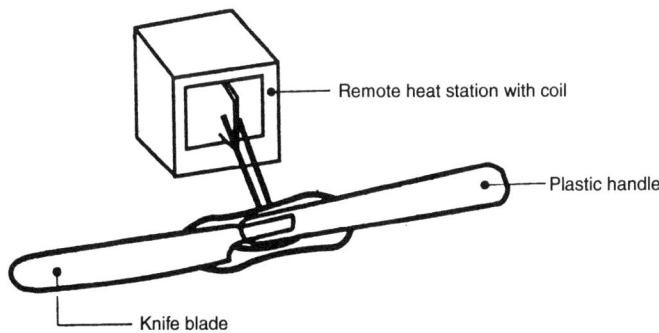

Fig. 4.53 Assembling a plastic handle onto a knife. Courtesy of Ameritherm Inc.

in 2 s prior to insertion in the glass-filled plastic connector housing. A 3 kW induction power supply delivers 1225 W to the part. Because the hole in the connector is a through hole rather than the usual blind hole, a guide rod can pass through the plastic housing to help guide and correctly locate the insert in the housing.

Metal Control Arms. Agriculture and yard care equipment has metal control arms with plastic handles. The plastic handles are often threaded onto the metal control arms. These handles eventually loosen and become a safety concern if a handle ends up missing or broken due to continuous adjustment. With induction heating, the molded handle and metal assembly is permanently fused together with a better torque strength than is achieved with the threaded technique.

During manufacture, each control arm/handle assembly is processed using the same coil and power supply with different heating times as necessary. A control arm is inserted into its corresponding handle, and the assembly is placed over the coil. Power is applied, and the handle is pressed onto the control arm as heating occurs. The steel control arm reaches a temperature of approximately 204 °C (400 °F), and heating time varies from 2 to 6 s, depending on the particular assembly.

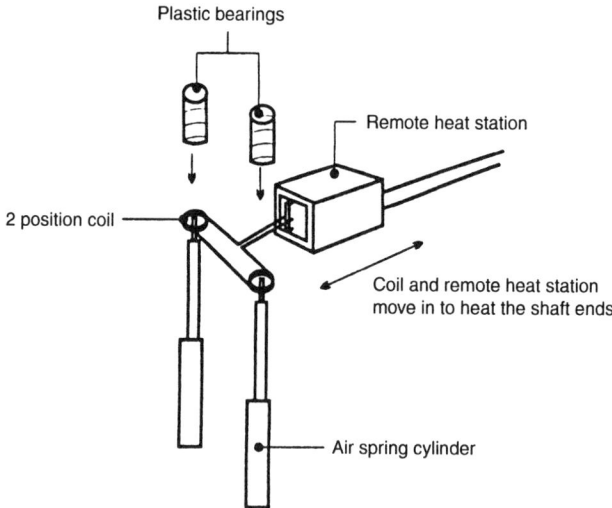

Fig. 4.54 Assembly of plastic bearing onto a steel shaft. Courtesy of Ameritherm Inc.

Inserting Large Brass Inserts. Modern automobile inlet manifolds are molded from glass-filled nylon and are bolted to the engine using 14.2 mm (9/16 in.) diameter brass inserts. These eight to ten large inserts are usually on one plane with a further number of smaller inserts on another plane to mount the exhaust gas recirculation (EGR) components and air duct or filter assemblies.

The 14.2 mm (9/16 in.) diameter brass inserts need 475 W of delivered power to achieve the required insert temperature of 371 °C (700 °F) in 4 s. Each insert is placed on a nonmagnetic stainless steel push rod, which is activated by an air cylinder. The insert is placed in the coil and heated for 4 s before being inserted into the manifold (Fig. 4.56, 4.57). The stainless steel push rod locates the insert in the manifold and passes through the coil. The heating cycle is only for the duration the insert is in the coil, and the power is switched off during the rest of the cycle so as not to heat the stainless push rod.

Multiple coils are required to heat the ten inserts simultaneously, and they are located in a heating plate through which the inserts pass. The coils can be part of a two or multiple position coil or be individually driven by 1 kW power supply. The smaller inserts can also be inserted at the same time with a separate coil and inserter push rod system. Typical cycle time is 10 to 15 s per manifold.

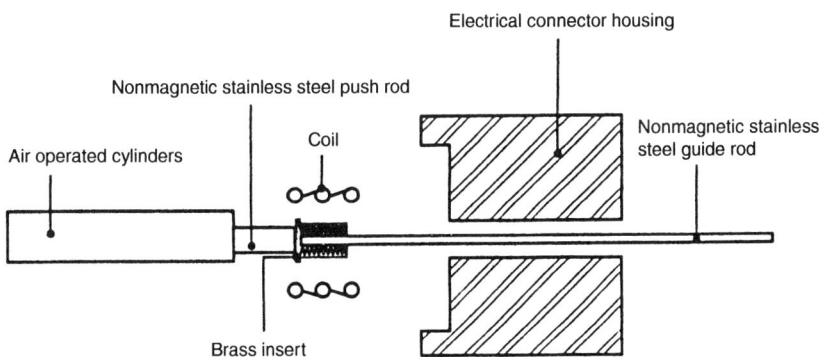

Fig. 4.55 Two-part electrical connector held together with a brass insert. Courtesy of Ameritherm Inc.

98 / Decoration and Assembly of Plastics

Fig. 4.56 Four-cylinder engine manifold showing brass inserts. Courtesy of Ameritherm Inc.

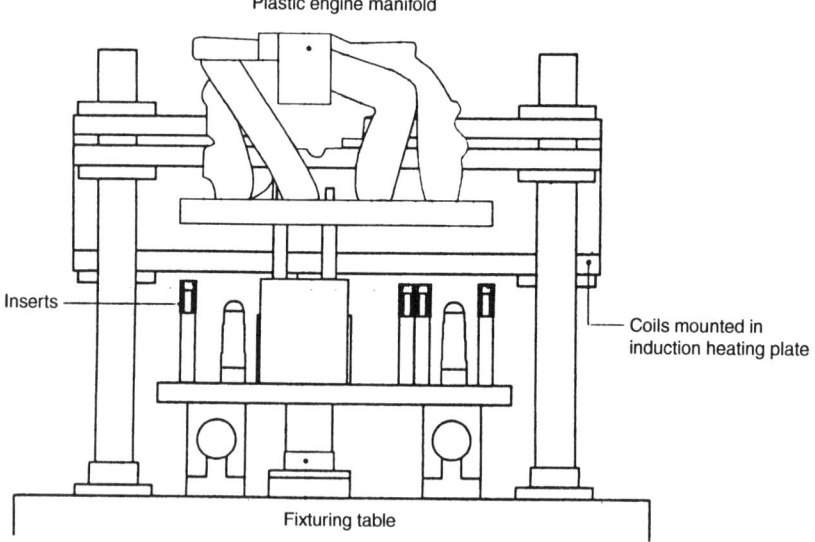

Fig. 4.57 Fixturing table for inserting inserts into engine manifold. Courtesy of Ameritherm Inc.

Induction Bonding

Induction bonding is a unique assembly process that lends itself to plastic product assembly. The principle is quite simple: an induced electric field is generated in proximity to a plastic part that is to be bonded to either metal or another plastic substrate (Fig. 4.58, 4.59).

A special adhesive compound containing 0.5 to 6.0% metallic particles is the bonding agent located between the two adherends. The induced electric

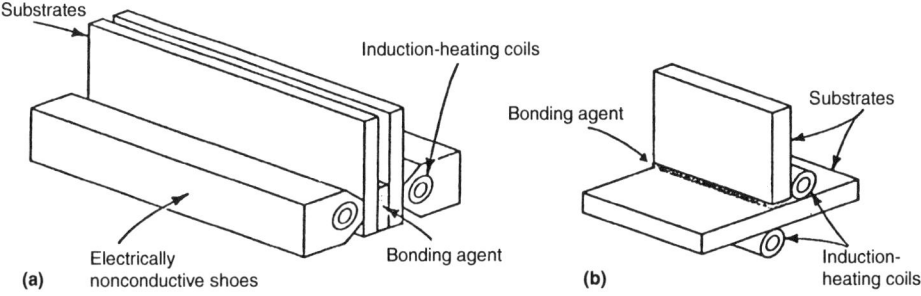

Fig. 4.58 (a) Cross section of basic setup for induction bonding. (b) Induction-heating bonding at a T-joint. The coils are placed diagonally above and below the joint. Source: Hellerbond Technology Company, Columbus, OH

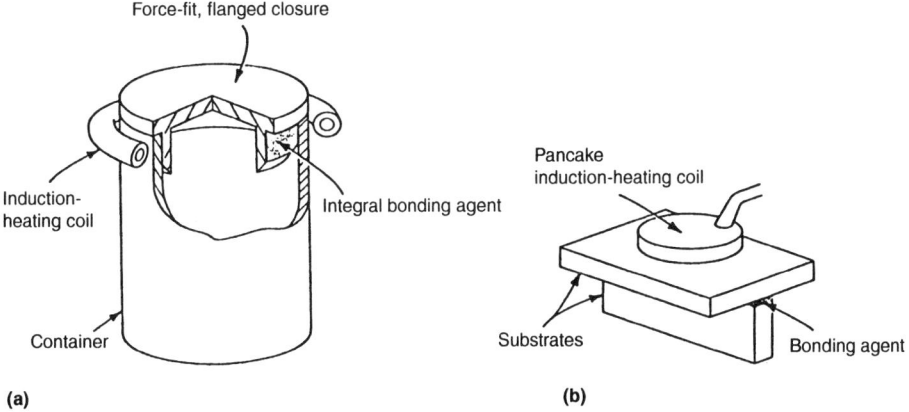

Fig. 4.59 (a) Integral bonding. Closure containing bonding agent is force-fit into container, and the two are subsequently bonded. (b) The pancake induction-heating coil is used when access to both sides of the bonding agent is unavailable. Source: Hellerbond Technology Company, Columbus, OH

field (2 to 6 MHz) will pass through the plastic with no effect, but when the adhesive is exposed to the field, the metallic particles are excited and generate thermal energy. This heat activates the otherwise dormant adhesive. Pressure is applied to the adherends, and the plastic is now bonded.

One advantage of induction bonding is that the adhesive system can be applied to the plastic part at the convenience of the manufacturer and activated at the convenience of the assembler. Another advantage is that the adhesive system can be made of the same resin as the plastic to be joined. Again, the presence of the metallic particle is required to effect the melting process, but because the same plastic is used, materials such as polyolefins and nylons can be bonded. Typically, these materials are difficult to bond using other adhesive systems.

Heat Staking

Heat staking is a method of assembling thermoplastics with other plastic, metal, or composite materials. During the heat staking process, a post is heated then reformed to form a head that mechanically retains the attachment (Fig. 4.60). Heat staking is accomplished by three principle methods: ultrasonic staking, hot air/cold staking, and direct contact with a heated probe.

Heat stakes are common in many assembly applications due to the quality and low assembly costs the process offers. In many consumer electronic devices, printed circuit board (PCB) and electromagnetic interference (EMI) shielding are mounted in plastic housings using heat stakes. Heat stakes are also widely used in automotive applications where they are used to assemble door panels and dashboards.

Heat staking was one of the original techniques used for assembling plastics. The early staking processes were similar to current methods with the exception of direct contact staking. The inconsistent softening characteristics (Fig. 4.61) sometimes caused the post material to stick to the staking probe. At other times, the same material would fracture because the staking process required a higher probe temperature. Another limitation for direct contact staking was that sophisticated equipment did not exist that could precisely control the probe temperature.

Modern Heat Staking: A Comparison of Three Methods

Ultrasonic Staking. The advances in materials and electronic controls, which have had a major impact on injection molding equipment in recent

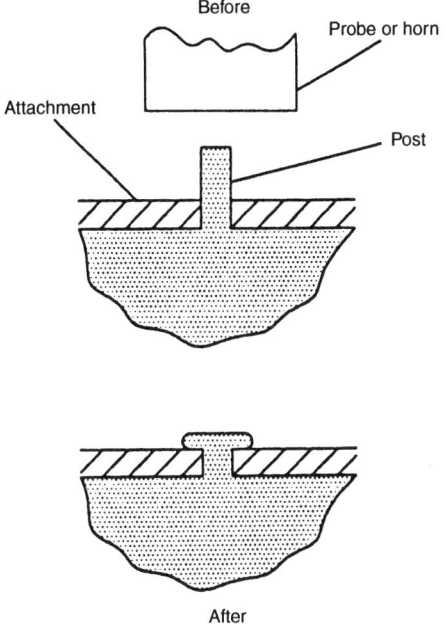

Fig. 4.60 Heat staking method. Courtesy of Young Technology

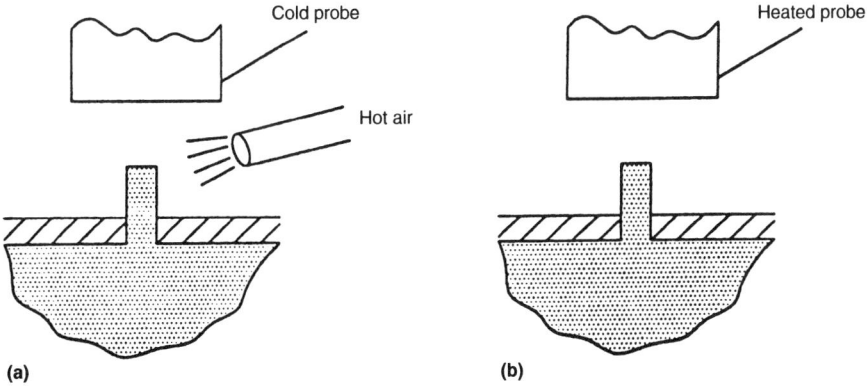

Fig. 4.61 Heat staking techniques of (a) early and (b) modern plastics. Courtesy of Young Technology

years, have also greatly improved the performance of heat staking equipment. Ultrasonic staking, which has been used for over thirty years to assemble plastics, is successful in many more demanding applications than previously possible due to the introduction of new, sophisticated microprocessor and computer process controls. The most advanced ultrasonic stakers now offer closed loop computer control of all process variables. Closed loop control enables the equipment to meet ISO 9000 and Food and Drug Administration (FDA) requirements for a process, which can be validated. These advanced control systems are not needed in all staking applications, but where consistency of the process is of paramount importance, the controls give users new opportunities for success.

Hot air/cold stake equipment, which is the most common method used for staking large components with any number of posts, provides a cost effective solution for many staking applications. The addition of more sophisticated controls to hot air/cold stake equipment provides a higher level of confidence in the system for users. However, some equipment manufacturers feel that the more sophisticated process controls produce very few real advantages for a technology that has proven itself and has changed little over time.

Direct Contact Staking. Another change in heat staking processes, which can be attributed to technology, is the reintroduction of the process called direct contact staking, or staking with a constant temperature probe. When hot air/cold stake and ultrasonic stake equipment were first developed, the materials and controllers necessary for the development of a precision direct contact heat staker were not available. In addition, before 1980, many thermoplastic materials had inconsistent melt characteristics. This inconsistency caused a direct contact heated probe to stake well sometimes and poorly at other times, which caused sticking and stringing that is unacceptable by current standards. Recently, special probe materials, bearing systems, and temperature controllers have been developed, which enable precise control of staking probe temperature and pressure. This control eliminates staking inconsistencies and the problem of the post material sticking to the probe. These changes have enabled the reintroduction of direct contact heat staking as a viable third method for heat staking of thermoplastics.

Table 4.7 lists features of the three staking methods and best demonstrates the similarities and differences. The comments were compiled from interviews with manufacturers and users of staking equipment. Table 4.7 provides a list of general characteristics of the three methods, but it does not address the many special process and equipment features available from

Table 4.7 Comparison of three staking methods

	Ultrasonic stake	Hot air/cold stake	Direct contact stake
Process description	An ultrasonic weld horn transmits high frequency vibrations that softens the post. The horn is simultaneously driven down on the post to form the stake	Hot air is used to soften the post and then a cold probe is driven down on the post to form the stake	A probe is simultaneously heated and driven down as the post softens to form the stake
Materials that are compatible to process	Any thermoplastic with less than 30% glass	Any thermoplastic with less than 30% glass	Any thermoplastic with less than 30% glass
Process variables	Frequency, amplitude, sonic time, hold time, trigger force, weld force	Hot air force, hot air direction, cold stake time, cold stake force, hot air time	Probe temperature, pressure time
Typical process dwell time	1 to 2 s	5 to 20 s	5 to 15 s
Number of posts that can be staked at one time	Limited to the number of posts that are in the horn contact area	No limitation	No limitation
Number of different height posts that can be staked at one time	Difference in post heights should be less than 12.7 to 25.4 mm ($\frac{1}{2}$ to 1 in.)	No limitation	No limitation
Can equipment be used for multiple applications by simple tooling changes?	Yes	No	Yes
Common reason for staking rejects	Damage to post at base caused by ultrasonic vibrations	Too much or not enough melting of post. Uneven head formation	Tightness if post and attachment are not properly designed
Maintenance requirements of equipment	Occasional resurfacing of horn face	Regular replacement of heater cartridges	Occasional resurfacing of probe face
Does the process effect materials or components close to the post?	Yes; vibrations can cause stress and damage to plastics and other materials near the post	Yes; hot air can damage or distort heat sensitive components near the post	Almost no effect from radiant heat on heat sensitive components
Does the process produce noise or heat uncomfortable for the operator?	Yes, noise can reach uncomfortable levels	Yes, hot air can cause the work area to become uncomfortably hot	No, only a manual amount of radiant heat is produced
Can the equipment be used for applications other than staking and swaging or thermoplastics?	Yes, welding and sealing of thermoplastics	No	Yes, seating of thermoplastics and any application requiring heat and pressure
Is the equipment custom built for each application, or is it a standard machine with custom tooling for each application?	Equipment is standard with custom tooling for each application	Equipment is custom built for most applications	Equipment is standard with custom tooling for each application
Time required to switch tooling for different applications	5 to 10 min	Equipment typically can not be used for more than one application	5 to 15 min
Are minor changes in post location difficult to accommodate after equipment and tooling are complete?	Depends on the type of change, horns can sometimes be modified or a new horn can be built	Some changes are easily accommodated, but some changes require major modifications to the equipment	Depends on the type of change, probes can be easily added, removed, or new looking can be built
Power requirements	Typically under 10 amps per machine	Typically 3 amps per post	Typically under 20 amps per machine for any number of posts

individual equipment manufacturers. Among the many equipment manufacturers and equipment users, a tremendous difference exists in the technology incorporated into the design of the equipment and the development of tooling, fixtures, and process parameters. Poorly designed equipment, tooling, fixtures, or process parameters can cause quality and consistency problems in an application that would otherwise be successful.

Advantages and Disadvantages

Heat staking provides many advantages when compared to other fastening techniques. For example, stakes can have an excellent cosmetic appearance and be used to assemble many consumer products where the stake head is visible. Stakes can be used to assemble any type of material to a thermoplastic part. With ultrasonic and direct contact staking methods, the attachment can even be heat sensitive, because these two methods subject the attachment to almost no heat. In addition, staking is a clean and fast process that can be done in any environment producing virtually zero rejects, because it is usually not operator dependent. It is also a process that lends itself to automation.

Staking can be used to assemble large or small components. Assemblies with 1 or 500 posts are staked with almost identical processes. There are few design restrictions on the location and number of posts that can be staked. When compared to screws, heat staking can provide the same type of mechanical bonding without many of the requirements of screw assembly. Staking does not require an inventory of materials, saving the cost of the screw and the time to manage the inventory. Posts for staking require less space than screw bosses and are typically 1.524 to 3.810 mm (0.060 to 0.150 in.) in diameter, where a screw boss is at least 2.540 mm (0.100 in.) in diameter. Heat staking adds no weight to an assembly, unlike screws and other fasteners. Screws limit the recyclability of the plastic components. In many applications, screws and stakes cannot be reworked, because a self-tapping screw does not provide the same holding strength if removed and replaced.

Heat stakes also have limitations, which should be considered when evaluating the suitability for a new application. Stakes can fail due to impact and fatigue. Because a stake provides only a point of attachment for two materials, it can be subjected to shear and tensile loads that will cause it to fail. A continuous weld provides a much higher strength assembly due to its ability to distribute loads over a larger area. An assembly that is heat staked together cannot produce a true hermetic seal unless a secondary seal is produced by a gasket or other means. A dust or particle resistant seal is possible with a well-designed heat staked assembly.

Post Features

A variety of post designs are shown in Fig. 4.62. A radiused post root adds strength to the post and minimizes molded-in stresses (Fig. 4.63). A common cause of rejects for heat-staked components is damage to the post prior to staking. During staking, especially ultrasonic staking, considerable forces are applied that can damage weak posts. The radius at the post root will strengthen the post to minimize rejects.

A uniform post top should be smooth to facilitate post material flow against a staking probe. An even flow of the post against the probe assures consistent head formation and the best cosmetic appearance. A radius or chamfer on the post top eases assembly of the retained part onto the post. A pointed post top should be avoided, because it can cause the post to deflect before the post material has softened. Deflection can lead to an uneven stake head or the post breaking at the root.

Relieved Post Root. If the attachment has a clearance hole that cannot be chamfered or radiused (e.g., PCB or sheet metal), a relieved post root design should be used. The relieved post root offers the benefits of the radiused post root but can be used with components without a chamfer or radiused clearance hole (Fig. 4.63b).

Stake Head

Dimensions. The diameter of the stake head should be larger than the clearance hole. Stake head material, which extends beyond the clearance hole, does not add to the strength or tightness of the stake (Fig. 4.64) and should be avoided due to the increase in process time required to form the excess material. The thickness of the stake head directly above the outer diameter of the post should be greater than one half the post diameter or greater than the boss wall thickness.

Stake Head Flush or Below Attachment Surface. A recess should be incorporated in the attachment to contain the stake head volume. A relief level is suggested to assure the stake head is below the surface (Fig. 4.64).

Stake Clearance. Components to be heat staked must be designed with adequate clearance for the stake and probe diameters (Fig. 4.65a). A clearance of >7.62 mm (>0.300 in.) is typically required for a 1.91 mm (0.075 in.) diameter post.

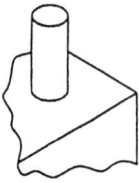

Advantages
- Post design is simple to mold
- Post is simple to modify
- Probe can have a simple flat tip

Disadvantages
- Sink marks are a problem if post diameter is too large
- Stakes do not provide rotational alignment for attachment

(a) Round solid posts

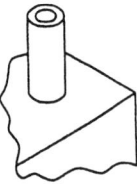

Advantages
- Probe can have a simple flat tip
- Post can also be used as a screw boss
- Post can be very strong

Disadvantages
- Stakes do not provide rotational alignment for attachment

(b) Round hollow posts

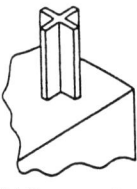

Advantages
- Post provides rotational alignment features for the attachment
- Post can be very strong
- Probe can have a simple flat tip

Disadvantages
- Post is more difficult to modify in the mold
- Appearance of the stake is not as good as with other post designs

(c) Cross posts

Advantages
- Rib provides rotational alignment features for the attachment
- Probe can have a simple flat tip

Disadvantages
- Rib can be difficult to stake using the hot air/cold staking method

(d) Ribs

Advantages
- Posts can be used to assemble parts with complex geometries

Disadvantages
- Post requires a complex probe face
- Alignment of the probe and post is critical for a good quality stake

(e) Angled posts

Fig. 4.62 Post designs used in hot staking and the advantages and disadvantages of each. Courtesy of Young Technology

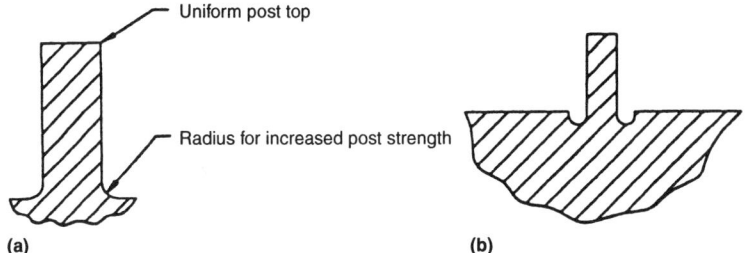

Fig. 4.63 (a) Radiused post root. (b) Relieved post root. Courtesy of Young Technology

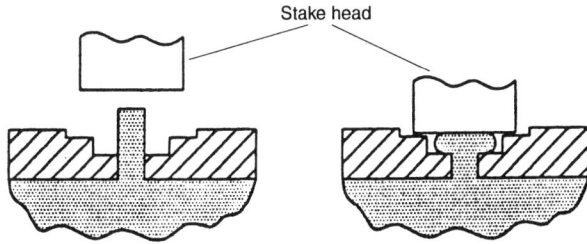

Fig. 4.64 Stake head that extends beyond the clearance hold. Courtesy of Young Technology

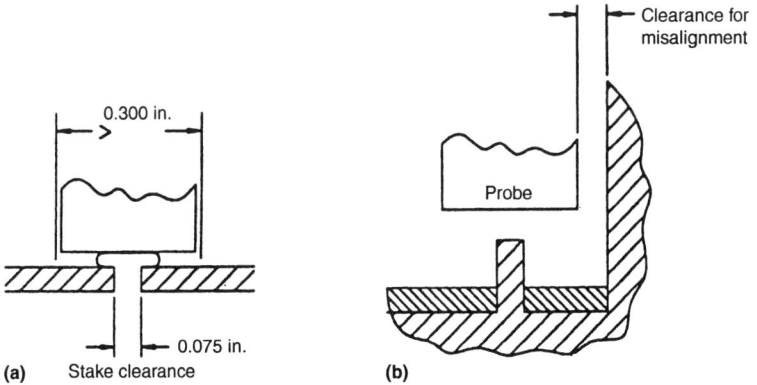

Fig. 4.65 (a) Stake clearance. (b) Probe clearance. Courtesy of Young Technology

Probe Clearance. The heated probe used for direct contact heat staking can be touching adjacent materials (Fig. 4.65b). There is typically little or no effect from the radiant heat of the probe on surrounding materials. If staking is done ultrasonically, the horn should not contact adjacent materials. If staking with the hot air/cold stake staking process, the staking probe can touch adjacent materials, but hot air may damage surrounding materials. With all methods, special consideration should be made to allow adequate clearance for the stake head and staking probe or horn due to misalignment. Misalignment of the probe or horn with the post of 0.127 to 0.508 mm (0.005 to 0.020 in.) is common due to variations in the molded part and clearance that is generally designed into the part holding fixtures.

The post clearance hole in the attachment should fit the post as tightly as possible (Fig. 4.66). This fit enables the stake to achieve maximum pull strength and assures that the attachment will be secured as tightly as possible. Excessively large clearance holes require the use of taller posts to create a stake head that can span the gap between the post and the attachment. Taller posts are also more prone to damage prior to staking and require a longer staking process. Maximum pull strength is achieved when the stake head and the interference fit of the post in the clearance hole are utilized to secure the attachment. The post clearance hole should also be chamfered or radiused to maximize the stake strength. The chamfer reduces the stress concentration factor on the stake head and post interface.

Calculating the Strength of a Stake. The tensile strength of a staked post equals the cross-sectional area of the post multiplied by the published tensile strength of the post material divided by the stress concentration factor. The 1.7 stress concentration factor is suitable for common clearance hole chamfers found in many injection molded attachments. It decreases as the radius of the post/head transfer region increases. For example, the tensile strength of a 1.524 mm (0.060 in.) diameter staked post is calculated: $(0.002827 \times 7000) / (1.7) = 11.6$ lbs.

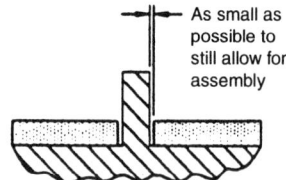

Fig. 4.66 Post clearance hole. Courtesy of Young Technology

Disassembly of Stakes with a Heated Probe. Disassembly for recycling or reworking of staked assemblies can be accomplished using precise heat and pressure (Fig. 4.67). Stakes can be removed by using a pointed probe, which upsets the stake material and separates the staked head from the post.

Disassembly of Stakes with Hot Air. Hot air can be blown directly on the stake head, softening it and causing the head to retract from the attachment. The head can then be removed with a cutting tool.

Reassembly. Components originally staked in assembly can be reused after disassembly. One method requires that the part be designed with twice the number of posts required for the first staking. The tooling is designed to stake only half the total number of posts. If the parts need to be disassembled by removing the staked posts, the same components can still be reassembled using the posts that remain unstaked. The second method incorporates one set of posts for heat staking, one set of bosses for screw attachment, and clearance holes in the attachment for both the posts and screws. During the original assembly operation, the posts are staked. When disassembled, the staked heads are removed, and the components are reassembled using screws.

Side Swaging

Attachments that cannot be manufactured with a post clearance hole can be assembled with thermoplastics by swaging a rib, post, or other feature in the plastic part over the attachment (Fig. 4.68). Swaging can also be used to form a cavity or relief in a component. In many situations, the feature created by swaging could have been molded into the part, but a secondary operation was required due to the complex shape of the feature.

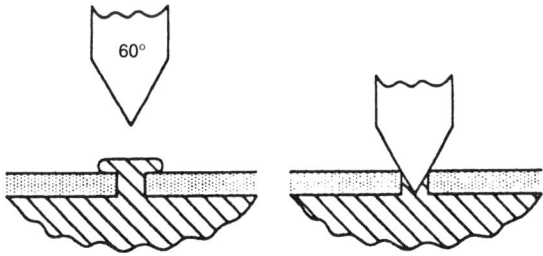

Fig. 4.67 Disassembly of stake by using a pointed probe. Courtesy of Young Technology

110 / Decoration and Assembly of Plastics

Common applications for swaging include attaching a glass display or lens to a plastic bezel. Other applications include securing O-rings, gaskets, washers, rods, or plastic components that are stock items and do not have post clearance holes.

The requirements for tightly swaging a component are the same as those for staking. The attachment should fit tightly against the feature in the plastic component that will be swaged.

Hot Gas Welding

Hot gas welding of plastics is a variation of metal welding techniques. The hot gas welding apparatus is quite simple (Fig. 4.69); it consists of a "gun" that acts as a source of hot air or inert gas. The hot gas welding system is frequently used to assemble and seal sheets of polypropylene in the fabrication of large chemical tanks. To accomplish the welding, the joint areas of the tank are designed to fit together while providing an area to add a bead of plastic melt, which acts as both adhesive and sealant. The hot gas welder is brought up to temperature, and air is allowed to pass over the heating

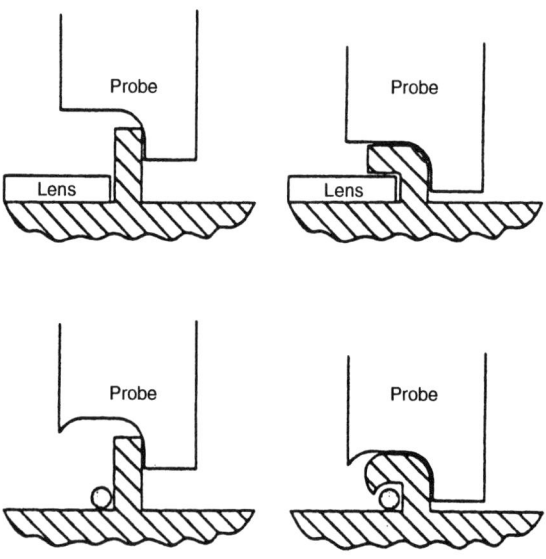

Fig. 4.68 Side swaging. Courtesy of Young Technology

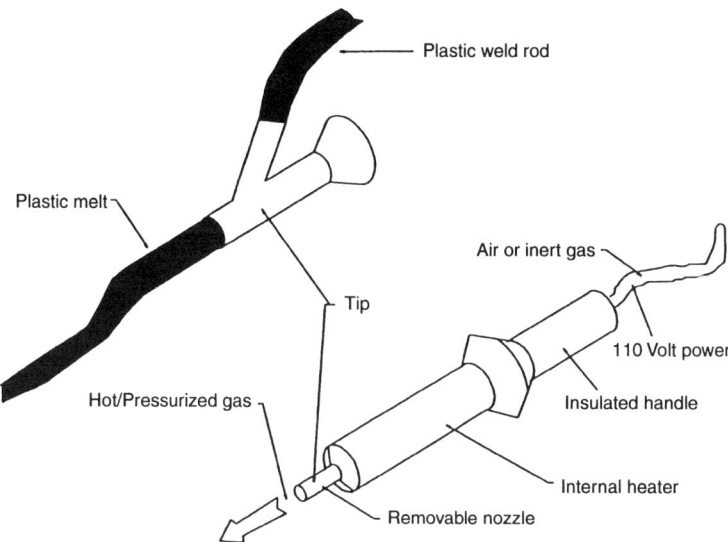

Fig. 4.69 Hot gas welding. Sheets of plastic are webbed together with a bond strength greater than the strength of the plastic itself

elements, creating a focused point of hot air. A weld rod made of the same plastic as the material to be joined is presented in front of the hot air stream and is allowed to melt and fit into the joint area. As the welding process continues, the bead is pulled in front of the hot air, thus allowing for a continuous process.

Recent advances in hot melt technology have resulted in the addition of pressure to the welding system, which provides a more uniform and consistent weld joint.

ACKNOWLEDGMENTS

The author wishes to express his gratitude to Sonics and Materials Inc., Ameritherm Inc., and Young Technology for contributions to this chapter.

5

Hot Stamping

Hot stamping is a plastics decoration and symbolization process that has been available to plastic part manufacturers for decades. Often considered a monochromatic (one color) process for simple-shaped plastic parts, many manufacturers overlook the hot stamping process. Today, new multicolor foils, diffraction surfaces, holographic imaging, and highly improved hot-stamp equipment demand a fresh look at this process. Product designers are rediscovering hot stamping as a solution to design challenges and as a means of avoiding environmental issues associated with other decorating processes.

Since the early 1950s, the rapid development of the plastics industry has brought about a high level of part performance, allowing plastics to make inroads in areas not thought possible a short time ago. The advances made in the hot stamping method have been just as dramatic, and the process is used in a wide range of markets, including automotive, appliance, cosmetic, electronic, furniture, housewares, medical, packaging, recreational, and toys (Fig. 5.1–5.3).

Advantages of Hot Stamping

The hot stamping process is still most widely used in decorating because of its convenience, versatility, and results. Because wet inks are not used, there are no offensive odors, environmental concerns, or storage problems. Furthermore, no ink mixing or messy clean up is needed, and the installation of a new color or design merely involves changing a roll of dry printed material, which minimizes set-up time.

Fig. 5.1 Hot stamped automotive parts

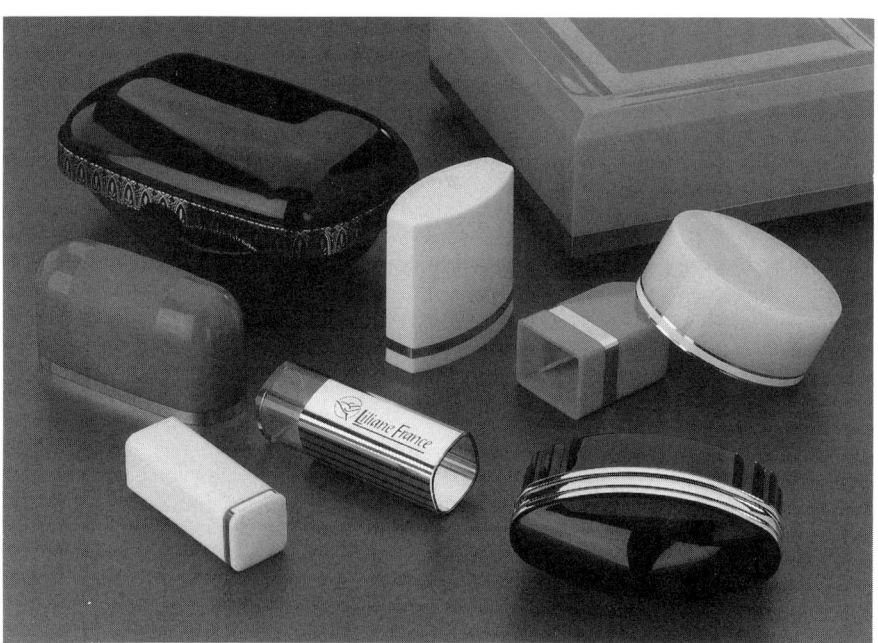

Fig. 5.2 Hot stamped cosmetic parts (noncircular)

Hot Stamping / 115

The versatility of hot stamping is one of its most important assets. Besides plastic materials, where emphasis is on thermoplastics, thermosets can also be decorated. In addition, leather, fabrics, paper products, coated wood, and prepainted metal are stamped with success. Hot stamping is also the only decorative method where permanent gold and silver metallic graphics can be produced. Foils are also manufactured in gloss or matte pigment colors, wood grain designs, brushed effects, and chromium for exterior use. Multicolored graphics can be accomplished with preprinted heat transfers and continuous patterned foils.

In addition to flat shapes, hot stamping foils can be applied to cylinders, slightly conical part shapes, and simple compound curves. Smooth finishes, textures, multilevel configurations, and raised graphics are also decorated with success.

Finally, the hot stamping process is capable of producing a high quality graphic, either small details or full coverage of large areas, that has excellent

Fig. 5.3 Hot stamped medical part

adhesion and abrasion resistance due to thermal bonding between the foil and the substrate. The permanent decoration is immediately dry to the touch and ready for handling and packaging (Fig. 5.4, 5.5).

Hot Stamping Process

To the casual observer, producing a permanent decoration on a part by the hot stamping process seems relatively quick and simple. In reality, if a typical machine cycle is periodically stopped and analyzed, there are a series of critical elements that must all take place to produce a quality graphic. A silicone rubber die is mounted to the heater head of the vertical acting machine and positioned directly over the decorating area of the part (Fig. 5.6). The die face is heated to a temperature near the melting point of the plastic substrate, usually in excess of 149 °C (300 °F). Suspended directly below the die in a path defined by two stripper bars is the hot-stamp foil. Typically, there is approximately a 12.7 mm (½ in.) space between the foil and the hot die face, so that the release characteristics of foil are

Fig. 5.4 Hot stamped beverage cases (product logos)

Hot Stamping / 117

Fig. 5.5 Hot stamped cassette (consumer electronics)

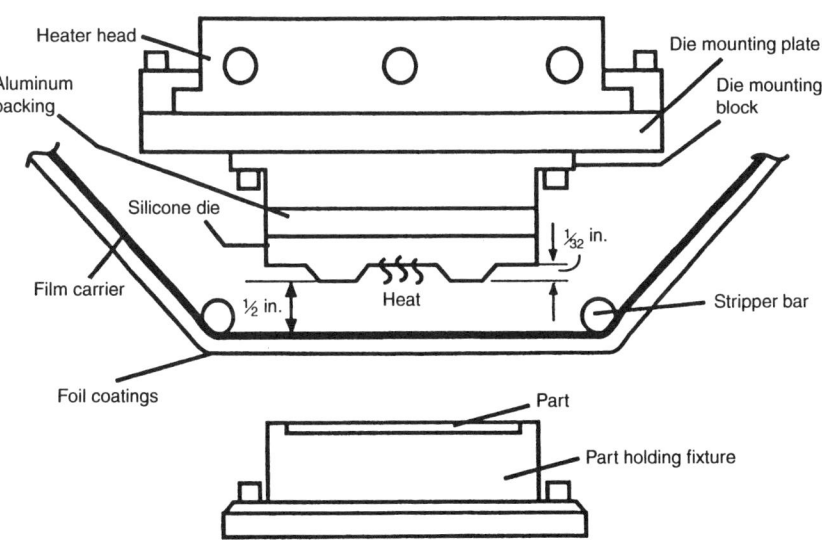

Fig. 5.6 Typical vertical stamping set-up. Courtesy of United Silicone, Inc.

affected as little as possible by heat convection. As an example of this, a common thermoplastic application is described in five distinct steps, and the key components are introduced as follows.

Step 1: Head Lowering. When the machine sequence has been initiated, the head will descend toward the decorating surface (Fig. 5.7). The foil, which travels with the head and at the same time maintains distance from the die face, contacts the part first, and the stripper bars act to push the foil tightly over the surface to remove any wrinkles.

Step 2: Die-to-Part Contact. A fraction of a second after the foil is pushed over the part, the hot-stamp die surface makes contact simultaneously with the foil and the rigidly supported plastic piece (Fig. 5.8). The pressure that is exerted accomplishes two things. First, the silicone rubber will compress and conform to any small surface variations in the part decorating area so that even contact is achieved. Second, the foil resins in the graphic area will break cleanly, creating a parting line.

Step 3: Dwell Time. During the dwell time, which is the time that the hot die is in contact with the foil and part, heat conduction will cause the release

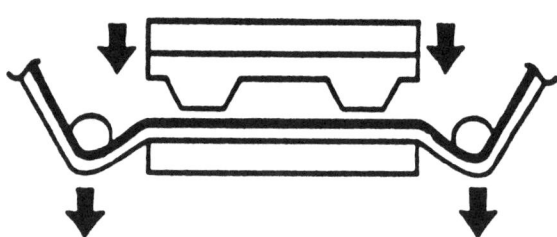

Fig. 5.7 Head lowering of vertical acting hot stamping machine. Courtesy of United Silicone, Inc.

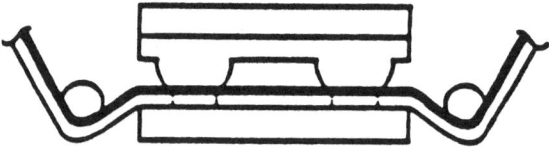

Fig. 5.8 Die-to-part contact of vertical acting hot stamping machine. Courtesy of United Silicone, Inc.

agents and resins of the foil coating to soften (Fig. 5.9). At the same time, the substrate surface will begin to soften, and the pressure exerted by the machine will help the resins penetrate the molten plastic to promote thermal bonding.

Step 4: Head Retraction. At the end of the dwell time, the head of the machine will retract and begin its ascent to the start position (Fig. 5.10). First, the hot stamping die lifts away from the substrate, while the foil remains on the surface for a split second, allowing the foil and part resins to begin to cool and harden. Then, as the die continues to rise, the foil will be peeled away, starting from the edges and working to the middle. At this point, the adhesion between the substrate and foil coatings is greater than the bond between the release agents in the coatings and the film carrier, resulting in a virtually complete deposition of the foil coatings.

Step 5: Foil Advance. As soon as the head of the machine returns to its original position, the foil will advance, and an unused section will be positioned under the die (Fig. 5.11). Finally, the hot stamped part can be removed from the fixture and handled or packaged as necessary, without any danger of rubbing the graphics off the substrate.

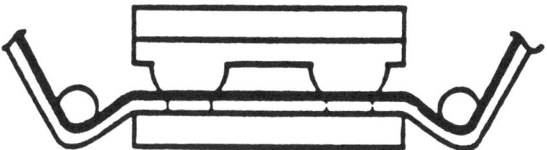

Fig. 5.9 Dwell time of vertical acting hot stamping machine. Courtesy of United Silicone, Inc.

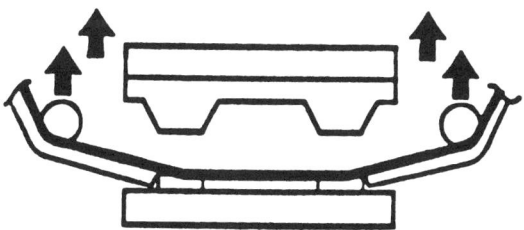

Fig. 5.10 Head retraction of vertical acting hot stamping machine. Courtesy of United Silicone, Inc.

Vertical Stamping Technology

Vertical acting applications, similar to those described in the section "Hot Stamping Process," are much more common than peripheral marking or roll-on decorating programs. Vertical stamping (Fig. 5.12) is ideal for applying foils or preprinted heat transfers to smaller areas of flat or slightly

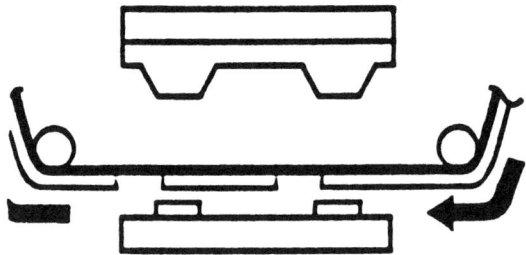

Fig. 5.11 Foil advance of vertical acting hot stamping machine. Courtesy of United Silicone, Inc.

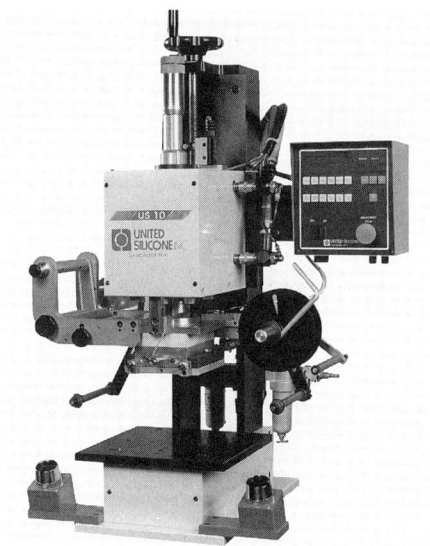

Fig. 5.12 Vertical hot stamping machine. Courtesy of United Silicone, Inc.

crowned parts and to a maximum of 90° on the circumference of cylindrical or slightly spherical-shaped parts (Fig. 5.13).

The major components of any vertical installation include the equipment, tooling, and hot-stamp foil. An example of this is an automotive tail light trim piece, where decorative bright silver graphics are stamped on the exterior surface. Like any application, all of the components involved must be properly selected, and the first task is to size the machine properly.

The vertical acting press has three main components: a rugged frame with fixture mounting table, a heated head with die mounting surface, and an automatic foil advancing device. When considering equipment size, it is important that there is enough frame clearance between the head and the machine table for the part to be decorated. Furthermore, the size of the head must exceed that of the die and be parallel to the table.

The most critical element is the amount of force that the machine can deliver. If a metal hot-stamp die is used, approximately one ton of force is required for 645 mm² (1 in.²) of contact area, while for a silicone rubber die, the desired output is about 400 lb/in.² (Fig. 5.14). With these force guidelines in mind, a vertical hot stamping machine should range in output from ½ to 20 tons. For applications less than 3 tons, the equipment is usually powered by a pneumatic cylinder, and a mechanical toggle configuration is common for a four to fifteen ton requirement. Similarly, an air/oil or hydraulic design can be utilized from four to twenty tons and beyond.

Surprisingly, in most vertical hot stamping applications, a single operator is involved in manually loading and unloading parts to a stationary fixture. Fixture access can be improved by incorporating a power slide table, and increased production output and convenience can be achieved by adding part transfer devices that permit loading parts while another part is being

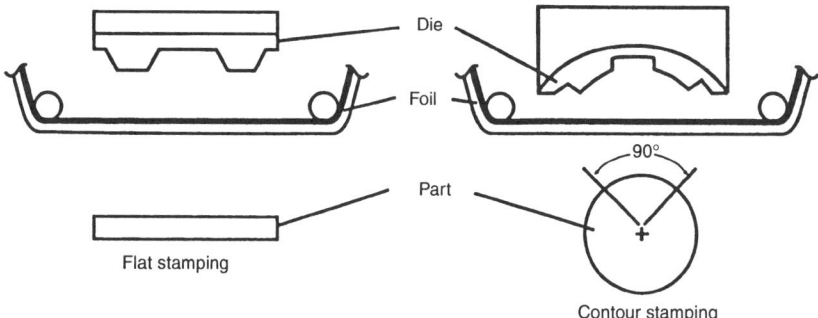

Fig. 5.13 Typical vertical stamping applications. Courtesy of United Silicone, Inc.

hot stamped. Common auxiliaries include rotary index tables or indexing conveyors. Labor content can be significantly reduced with the addition of automatic feeding and discharge equipment. Vibratory bowl feeders and vibratory stacker feeders are common, and pneumatically activated units and pick-and-place mechanisms are the most popular for part removal.

Once the equipment is selected, attention can be focused on the custom tooling set. Often during discussions on this subject, emphasis is placed on the part-holding fixture details. However, it should be noted that a fixture only comprises half of the set. The other half of the equation, the hot stamping die, is equally important and should be correctly mated to the part on the fixture in order to function properly. The components should be manufactured together to guarantee a good fit, especially with a contoured application.

The various components of a fixture should be constructed directly from the parts, not from a part print, and four key elements should be considered:

- *The decorating surface* should be presented as parallel as possible to the head of the machine.
- *The fixture* should hold the part firmly in place throughout the stamping cycle, and it should be designed so that part-to-part location is consistent.

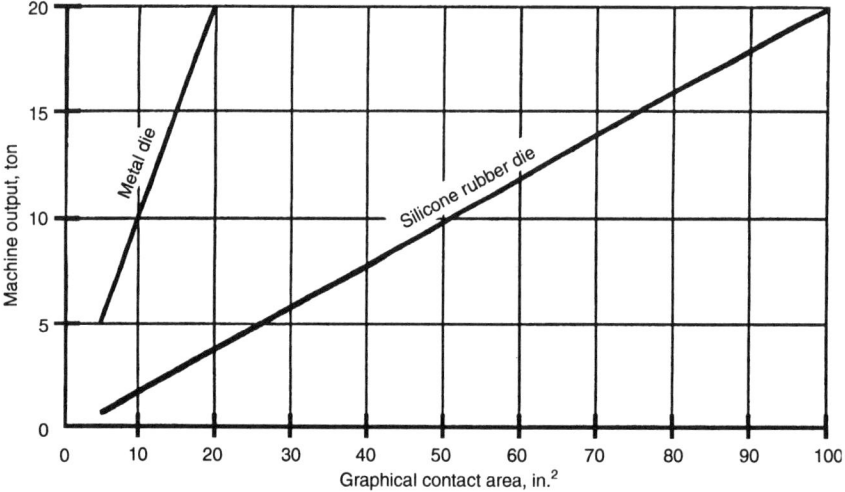

Fig. 5.14 Vertical hot stamping machine selection guide. Courtesy of United Silicone, Inc.

- *Support* behind the decorating surface is essential regardless of how rigid the plastic part seems. Even minimal part or fixture deflection can lead to decoration quality problems.
- *Convenient operator access* for part loading and unloading purposes should be present.

Fixturing components are produced in a variety of materials, including aluminum, steel, polyvinyl chloride (PVC), nylon, Teflon (E.I. Du Pont de Nemours & Co., Inc., Wilmington, DE), and a range of rigid and resilient casting compounds. For part-holding fixture base plates, aluminum is selected over steel, because it is lighter but still durable, is easy to machine, and will not rust. Although aluminum is also popular for part supports, when clear or high-gloss substrates are decorated, PVC is recommended. As long as the material is kept clean, plastics will not be scuffed. Nylon and Teflon are like PVC in this way, but they are also ideal for automatic part discharge applications because of a low coefficient of friction. Finally, for many contoured applications, a coating of urethane is cast to the shape of the part and bonded to a metal base. This addition gives the advantage of a tighter fit than is provided by any other material machined to the shape.

As stated earlier, the fixture must be properly mated to the shape of the die face. However, no matter how good the match is, minor adjustments are always required during set-up. Some common features include oversize mounting holes or slots for positioning adjustment and jack screws in the corners for leveling adjustment.

Specifications for the hot-stamp die, the second component of a custom tool set, are selected based on one of two decorating part-surface configurations: flush stamping and tipping (Fig. 5.15). For flush stamping applications, the graphics to be decorated are raised on the die surface a minimum

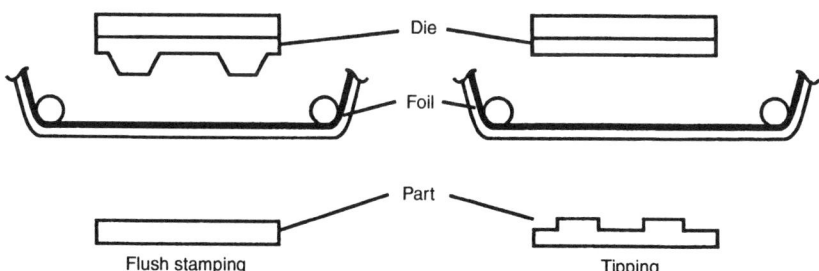

Fig. 5.15 Typical part surface configurations and die selections. Courtesy of United Silicone, Inc.

of 0.794 mm ($\frac{1}{32}$ in.), and molded silicone rubber dies are the material of choice in most plastics applications. Custom silicone formulations are available in various levels of hardness (durometer), which permit the material to conform to the surface variations in plastic molded parts while maintaining excellent graphic detail. The silicone rubber, which has outstanding heat stability and bondability properties, must be bonded to a metal backing, usually aluminum, for structural as well as thermal conductivity purposes. The dies are available in flat, multilevel, or contoured configurations. The latter can be designed to bend to the shape of a die-mounting block or molded directly from a contoured mold and bonded to a curved foundation.

Although molded silicone rubber dies are extremely popular for hot stamping on plastics (Fig. 5.16), there are certain situations where metal dies provide advantages. Like their rubber counterparts, the dies are available flat or contoured, and hardened steel dies are the most utilized because of their longevity. Other materials, including brass, copper, and magnesium, which are softer and less expensive than steel, are ideal for shorter production runs. The metal die tends to embed the foil into the plastic surface, which provides natural abrasion protection but a less consistent decoration than a rubber die.

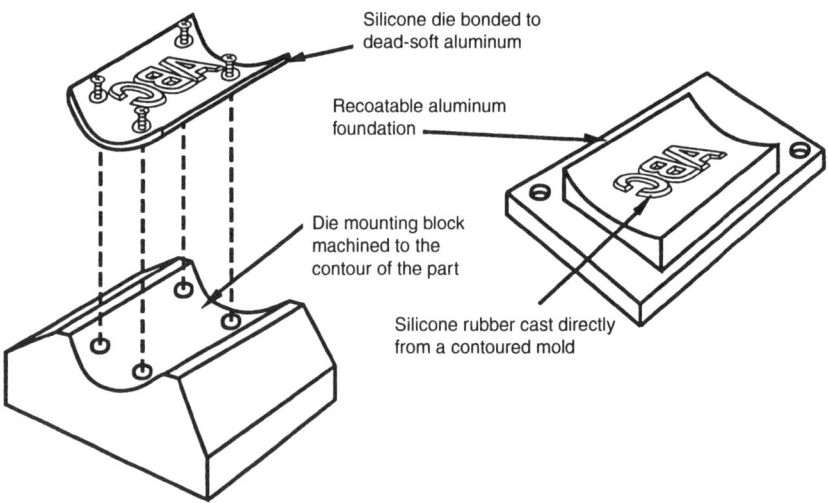

Fig. 5.16 Contoured silicone rubber die options. Courtesy of United Silicone, Inc.

For tipping applications, the die material choice is limited to silicone rubber. A sheet stock configuration is most popular, and it is commonly used for highlighting raised graphics and beads and is available in thousands of combinations of hardness, size, and thickness. Also, it may be utilized flat or can be bent to fit workpiece contours (Fig. 5.17).

Silicone Rubber Sheets and Dies. Silicone rubber tools, which conform to the surface variations inherent in plastic molded parts, are the material of choice in decorating applications. United Silicone, Inc. (Lancaster, PA) offers custom formulations with balanced properties for superior high-heat performance. Their proprietary rubber-to-metal bonding systems encourage long life, and rigid process controls assure consistent product quality (Tables 5.1–5.6).

Silicone rubber sheet stock is commonly used for tipping raised graphics and beads or for applying multicolor heat transfers. It is available in thousands of combinations of material, size, and thickness for virtually any application (Fig. 5.18). Silicone rubber is carefully molded, cured, and

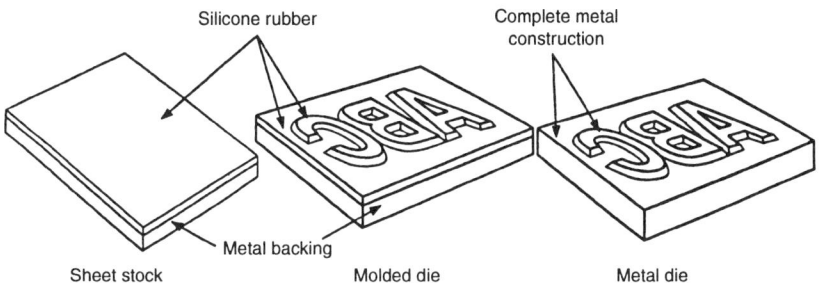

Fig. 5.17 Hot stamping die comparison. Courtesy of United Silicone, Inc.

Table 5.1 Silicone rubber product comparison

Formulation	Description	Application
Supersil (red color)	Excellent general-purpose material specifically formulated for fast turnaround	Conventional vertical operations requiring moderate temperatures and cycle times; typically manually fed
Ultrasil (red color)	Premium, high-performance material with outstanding rubber-to-metal bondability	Demanding vertical operations requiring high temperatures and/or high pressures; usually manually fed
Thermosil (dark brown color)	Advanced, high-performance material with enhanced heat recovery properties	Semiautomatic and automatic vertical operations requiring very stable high-temperatures and very rapid recovery

Courtesy of United Silicone, Inc.

precision ground for quality and performance (Fig. 5.19). It can be utilized flat, or it can be bent to fit workpiece contours. Custom molded dies (Table 5.7), which are popular for hot stamping graphics on flat or contoured parts, are formed from silicone rubber. Full graphic art and photographic capability assure faithful art reproduction when using silicone rubber. It is available in flat, multilevel, or contoured configurations for diverse applications.

Table 5.2 Silicone rubber product specifications

Specification(a)	Sheets	Molded dies
Rubber durometer (hardness)(b), Scale A	40–90(c)	40–90
Dual durometer(b)	Available	Available
Texture(b)	Available	Available
Size, mm (in.)	304.8 × 304.8 to 914.4 × 1219.2 (12 × 12 to 36 × 48)(d)	Up to 558.8 × 635 (22 × 25)
Raised graphics, mm (in.)	Not available	0.794 or 1.587 ($1/32$ or $1/16$) standard
Silicone thickness(b), mm (in.)	0.794–12.7 ($1/32$–$1/2$)(e)	0.794–4.762 ($1/32$–$3/16$)(e)
Metal backing(b), mm (in.)		
Half-hard aluminum (standard)	0.397–12.7 ($1/64$–$1/2$)	0.397–12.7 ($1/64$–$1/2$)
Dead-soft aluminum (bendable)	0.397–1.587 ($1/64$–$1/16$)	0.397–1.587 ($1/64$–$1/16$)
Steel	0.397–25.4 ($1/64$–1)	0.397–25.4 ($1/64$–1)
Brass	0.794–25.4 ($1/32$–1)	0.794–25.4 ($1/32$–1)
Contact surface shapes	Bi-level (surface ground)	Bi-level, crown(b), or contour

(a) For information on rubber formulation, see Table 5.1. (b) For more information, see Table 5.3. (c) 80 durometer is standard. (d) 304.8 × 304.8 mm (12 × 12 in.) is most common. (e) 3.175 mm ($1/8$ in.) is most common. Courtesy of United Silicone, Inc.

Table 5.3 Notes on specifications of silicone rubber

Specification	Note
Rubber durometer	Measure of material hardness as expressed on the Shore A scale. Low-durometer rubber exhibits high-elongation properties and improves coverage on irregular surfaces. Rubber in higher durometer ranges is more tear resistant and will withstand pressure with minimum distortion.
Dual durometer	Two different rubber durometers are molded together in levels. Usually a harder material is used in the face and a softer one in the background. This combination provides a soft pliable belly beneath the printing surface to help absorb variations in part wall thickness.
Texture in contact surface	For vertical applications with large surface areas, a texture in the rubber face will help redistribute trapped air.
Silicone thickness	The thicker the material is, the better the die life will be. However, the heat transfer is reduced.
Crown in contact surface	For vertical applications with large surface areas, a convex crown (as little as 2.54 mm, or 0.01 in.) in the rubber face will help eliminate air entrapment.
Metal backing	Silicone rubber bonds much better to aluminum than steel or brass. The aluminum is more porous than the other materials and effectively provides a larger bonding area.

Courtesy of United Silicone, Inc.

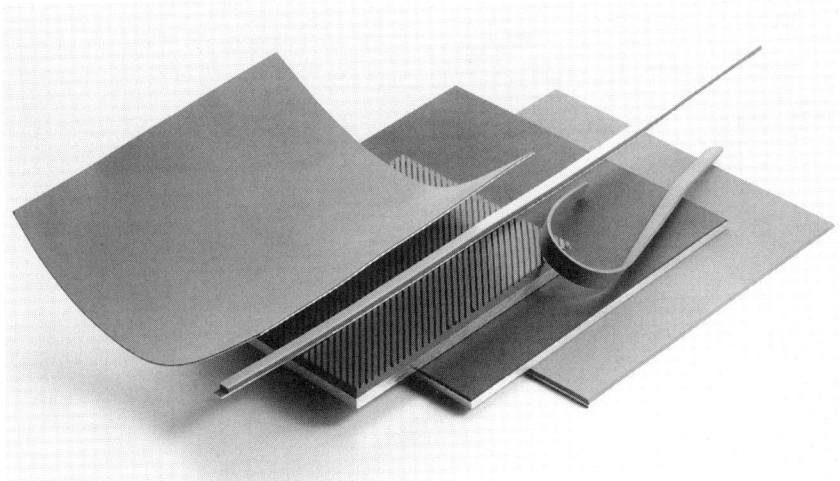

Fig. 5.18 Sheet stock for hot stamping (silicone rubber)

Table 5.4 Common causes for silicone die failure

Cause	Description	Symptom
Over compression	Over compression of the die causes lateral forces at the bond, which leads to delamination.	Delamination, with black or gray powder residue at the bond line
Excessive heat	Rubber materials can withstand service temperatures over 316 °C (600 °F). However, die life is shortened dramatically by exceeding practical limits. Often, using thinner rubber allows the platen temperature to be lowered as much as 56 °C (100 °F) and still achieve the same die face temperature. Thinner rubber also allows faster heat recovery times between cycles.	Rubber is brittle or appears dried-out. When temperatures over 316 °C (600 °F) are used, delamination may occur.
Compression set	Stamping on narrow raised areas, such as beads and lettering, ultimately causes the rubber die or pad to take a permanent set. When this condition becomes extreme, stamping quality declines.	Depressions in the rubber contact surface
Cutting and tearing	Excessive pressure, sharp part edges, and/or misloaded parts are the most common causes. Avoiding these conditions improves die life.	Fractures in the rubber surface
Thermal shock	Extreme and rapid temperature change can cause thermal expansion beyond the bond capability of the material. Many users leave hot stamping machines on "after hours" to avoid constant temperature changes, which also reduces start-up time at the next production period.	Rubber is brittle or appears dried-out. Partial delamination may occur.

Courtesy of United Silicone, Inc.

Table 5.5 Properties of silicone rubber formulations

Durometer(a), Shore A	Tensile strength(b) kPa	Tensile strength(b) psi	Elongation(c), %	Compression set(d), %	Heat resistance(e) °C	Heat resistance(e) °F	Specific gravity(f)
Supersil (red color)							
90	5206	755	140	14	260	500	1.77
80	5240	760	140	15	260	500	1.73
70	5033	730	180	12	260	500	1.67
60	5240	760	250	13	260	500	1.57
50	5723	830	440	15	260	500	1.45
40	7308	1060	585	10	260	500	1.26
Ultrasil (red color)							
90	6757	980	60	30	316	600	1.74
80	7446	1080	120	24	316	600	1.73
70	6895	1000	170	20	316	600	1.57
60	6964	1010	240	18	316	600	1.45
50	6274	910	380	17	316	600	1.36
40	5654	820	520	15	316	600	1.12
Thermosil (brown color)							
90	5378	780	95	30	316	600	2.11
80	6205	900	130	24	316	600	2.03
70	5550	805	220	24	316	600	1.84
60	4895	710	310	25	316	600	1.64

(a) Durometer is the hardness of a material as measured on the Shore A scale (90 durometer = hardest). (b) Tensile strength is the pulling stress just before a material breaks into two pieces (7446 kPa, or 1080 psi, is strongest). (c) Elongation is the fractional increase in length of a material stressed in tension just before it breaks into two pieces (580% = most elastic). (d) Compression set is the measure of material resiliency after being subjected to compression and heat (10% = most resilient). (e) Heat resistance is the ability of a material to remain bonded to metal during exposure to extreme temperature (316 °C, or 600 °F, is most resistant). (f) Specific gravity is the density of a material divided by that of water (2.11 = best thermal conductivity). Courtesy of United Silicone, Inc.

Table 5.6 Silicone rubber heat loss

Station thickness, mm (in.)	Temperature reading(a), °C (°F)	Temperature difference(b), °C (°F)	Difference, %
0.254 (0.010)	182 (360)	22 (40)	10
0.508 (0.020)	177 (350)	27 (50)	12.5
0.813 (0.032)	166 (330)	38 (70)	17.5
1.58 (0.062)	154 (310)	50 (90)	22.5
3.18 (0.125)	143 (290)	61 (110)	27.5
6.35 (0.250)	132 (270)	72 (130)	32.5
9.53 (0.375)	121 (250)	83 (150)	37.5
12.7 (0.500)	110 (230)	94 (170)	42.5
15.88 (0.625)	104 (220)	100 (180)	45
19.05 (0.750)	99 (210)	105 (190)	47.5
22.2 (0.875)	93 (200)	111 (200)	50
25.4 (1.0)	88 (190)	116 (210)	52.5

(a) Temperature readings taken to nearest 6 °C (10 °F). (b) Platen temperature, 204 °C (400 °F). Temperature differences between silicone die surface and platen would be greater if heat was absorbed by parts being decorated and would depend on stamping rates and part temperature. Courtesy of United Silicone, Inc.

Hot Stamping Foils

The final component of any vertical application, the hot stamping foil, is available in an infinite variety of colors and designs. The most common foils are metallics and pigments, but wood grain designs, brushed effects, continuous patterns, and chromium (for exterior use) are also used extensively.

Regardless of the type of finish, all foils consist of various thin layers deposited on a film carrier. The main function of the carrier is to deliver the print medium to the hot stamping area, and it should be able to withstand the heat of the operation without breaking or distorting. The material, typically polyester in ½, ¾, or 1 mil thicknesses, also acts as a barrier from the heated die to protect the release characteristics of the foil.

Recent advances in foil technology have seen the development of diffraction foils, which allow multiple light refraction, and holographic foils,

Fig. 5.19 Hot stamping supplies (dies and pad)

which allow three-dimensional (3-D) effects. The 3-D holographic foils have found a niche in security validation of credit cards and software products. Both diffraction and holographic foils are seeing increased usage in the decorative markets.

Carrier thickness selection is based on the hot-stamp applications and/or foil coating configuration. For example, because all plastic films react to heat by wrinkling, a thicker film is used for an application that requires a lengthy dwell time and/or an excessive die surface temperature. Furthermore, certain foils with many layers of thick coatings, such as woodgrain designs, require a thicker carrier to support the extra weight.

Metallic Foil Construction

Figure 5.20 shows the metallic foil construction process and components.

The release coat is usually a wax coat that melts at a specific temperature to release the decorative coatings from the carrier.

Table 5.7 Comparison of silicone dies to metal dies

Category	Silicone rubber dies	Metal dies
Materials available	Supersil, Ultrasil, Thermosil	Magnesium, copper, brass, steel
Average die life	Contoured, X; flat, 2X	Magnesium, Y; copper, 4Y; brass, 18Y; steel, 20Y+
Compensate for surface variations (sinks, etc.)	Usually	No
Compensate for variation in material thickness	Usually	No
Set-up time	Fast process	Lengthy process
Approximate pressure required	350 lb/in.^{2}	2000 lb/in.^{2}
Approximate thermostat temperature required	56 °C (100 °F) more than metal (~177–204 °C, or ~350–400 °F)	56 °C (100 °F) less than rubber (~121–149 °C, or ~250–300 °F)
Approximate dwell time required	50% more than metal	30% less than rubber
Decorative finish	Usually foil lays on top of plastic surface, more consistent decoration	Die embeds foil into plastic surface, less consistent decoration
Price		
Makeready	Separate charge	No
Initial die	Less expensive than metal	More expensive than rubber(a)
Spare die	Same cost as initial die	Same cost as initial die
Acceptable thermoplastic substrates	ABS, acetal, acrylic, polyamid (nylon), polycarbonate, polyethylene, polystyrene, PVC (plasticized)(b), PVC (rigid), SAN(b), ultraviolet coating	ABS, acetal(b), acrylic, polyamid (nylon)(b), polycarbonate(b), polyethylene(b), polystyrene, PVC (plasticized), PVC (rigid), SAN(b), ultraviolet coating
Acceptable thermoset substrates	Epoxy/epoxy coating, polyurethane	Epoxy/epoxy coating, phenolic, polyurethane(b)

(a) Magnesium, $X; copper, $2X; brass $8X; steel $10X. (b) Preferred die. Courtesy of United Silicone, Inc.

Protective Lacquer. Once the print medium is released from the carrier, this layer is the top coating on the decorated substrate, providing abrasion and chemical resistance. The lacquer also defines color. For example, a transparent amber-tinted lacquer is used for gold foil.

Metallizing. The vacuum metallizing process is used to provide the mirrored finish of a metallic foil.

Size. The purpose of this coating, which consists of resins matching those in the substrate to be decorated, is to bond the stamping foil layers to the part. For example, to stamp polypropylene, an adhesive is used that fuses to polypropylene.

Pigment Foil Construction

The most common pigment foils are made in matte, gloss, or day-glo finishes.

Matte Finish. The composition of matte foils include colored pigment particles, fillers, extenders, resins, and a plasticizer to bind the materials to the film carrier (Fig. 5.21). Unlike metallic foils, which have four distinct coatings, matte foils have all of the materials mixed in a single coating. With this configuration, the particles separate from each other as the foil strips from the plastic part leaving some residue on the carrier. Disadvantages of matte foils include that they scratch easily and can smear.

Gloss Finish. The composition of gloss foils include two layers (Fig. 5.22). The first layer is very similar to that of the matte finish, except that more resins are added that make the finish glossy and tighten the release. The

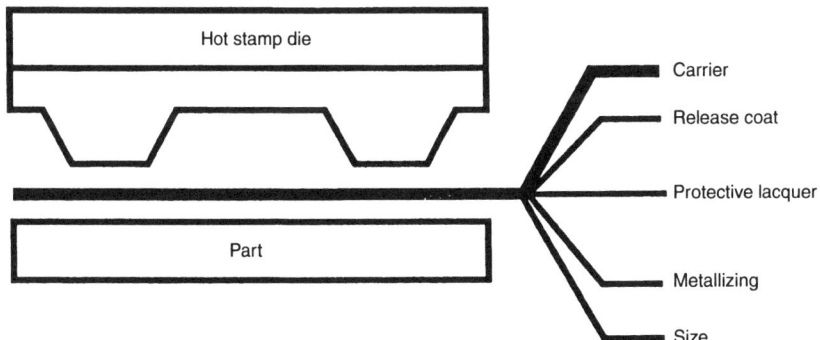

Fig. 5.20 Metallic foil construction. Courtesy of United Silicone, Inc.

additional layer is used to help release the particles from the film carrier onto the part and, at the same time, protect the stamp from marring and smearing.

Day-glo finish foils are composed of three layers (Fig. 5.23). The first layer is very similar to that of the matte finish, except that more pigment particles are added to improve opacity. The back-up layer comes next, and it is formulated like the first, but the color is white to provide additional opacity so dark part colors will not show. The final layer is used to help release all the particles from the film carrier onto the part.

Peripheral Marking Technology

Peripheral marking is mainly used for applying foils or preprinted heat transfers to the periphery of cylindrical, as well as slightly conical, parts. The advantage of this process is that up to 360° of the part circumference

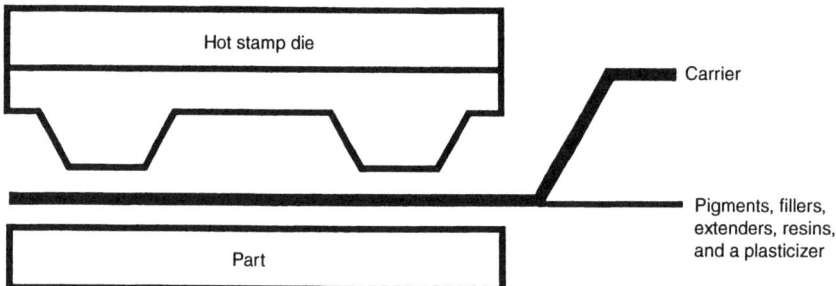

Fig. 5.21 Matte pigment foil construction. Courtesy of United Silicone, Inc.

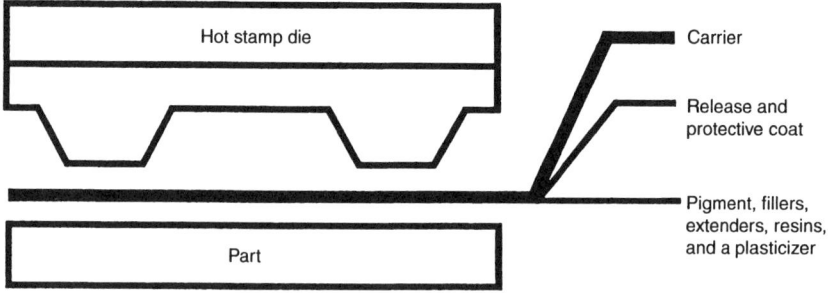

Fig. 5.22 Gloss pigment foil construction. Courtesy of United Silicone, Inc.

can be decorated in one machine cycle. An example of this is a plastic lipstick tube where silver or gold metallic stripes are applied to produce a continuous design, which can be completed in one of two ways. With the moving part method, the piece is rolled under a heated hot-stamp die with raised graphics while the tube traverses from side-to-side. On the other hand, the same tube remains static with the stationary part method while it is rotated by a heated silicone rubber roller (Fig. 5.24).

The upper half of the moving part peripheral marking machine is very similar to the vertical acting concept. Mounted to a rugged frame are the heater head and foil feed, which are sized appropriately to exceed the dimensions of the hot-stamp die. The stamping force is typically developed

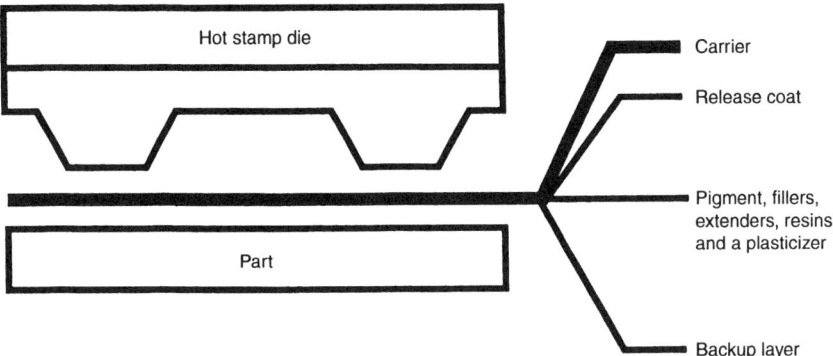

Fig. 5.23 Day-glo pigment foil construction. Courtesy of United Silicone, Inc.

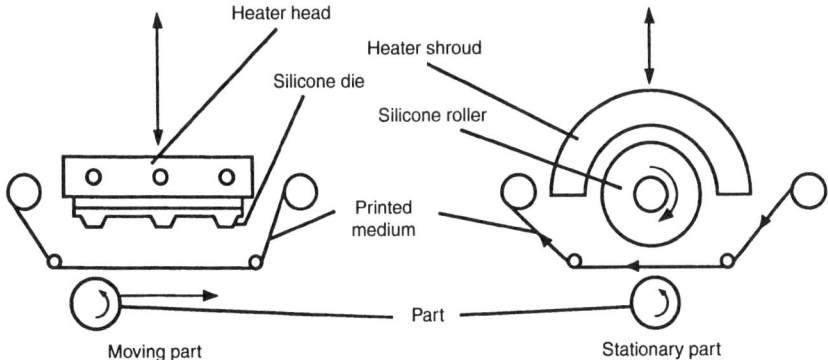

Fig. 5.24 Common peripheral marking methods. Courtesy of United Silicone, Inc.

with a double-ended pneumatic cylinder. Because of the off-center loading, as the head width increases, guide rods in linear bushings are incorporated for stability at the edges, while part tapers necessitate the employment of an angular adjustment for the head.

The lower half of the machine is configured in one of two ways, depending on the desired production output. The highest efficiency is obtained by utilizing a variable speed, electric motor driven, over/under type chain conveyor to facilitate rapid, continuous part feed. On the other hand, a reciprocating shuttle design provides adequate output. Quality decorations are possible, because an air cylinder with hydraulic dampening in the feed direction produces smooth and consistent part transfer. Either concept can be integrated with automatic feed and ejection devices for production optimization.

Flat hot stamping dies are used with moving part peripheral marking machines so that letters and numerals can be decorated in conjunction with decorative stripes. Therefore, tooling and foil technology is very similar to that of a vertical acting machine. However, there are some special requirements. Due to the part transfer speed, fast heat recovery between decorations at the die surface is at a premium. Hardened steel, brass, copper, or magnesium dies are popular because of this recovery, but silicone rubber dies are often specified for plastic applications when the metal die is incapable of producing a consistent hot stamp because of part variance. With silicone, acceptable heat recovery rates are obtained when the rubber thickness is minimized, and heat retention behind the die surface is maximized with the use of a steel backing and mounting block.

Cradle and mandrel type part holding fixtures are the most common for cylindrical parts. Both designs incorporate low-friction ball bearings to allow for smooth part rotation. A cradle concept can be specified when the part is solid or strong enough to support itself without deflection, such as a tube. When the graphic does not contain a continuous element or when orientation of the art to a specific location on the cylinder is critical (Fig. 5.25), the mandrel concept is often interfaced with a rack-and-pinion to drive the part during decoration.

Silver and gold metallic foils are the most common for peripheral marking applications, but various pigment designs are also used. Foil construction, although usually identical to that used in vertical applications, is sometimes modified when fast decorating speeds and/or small, high detail graphics are involved. Typically, additional size coats are added to improve the bond with the metallized layer and, at the same time, maximize the adhesion of the foil to the substrate. The stationary part peripheral marking method commonly utilizes the same foil configuration and part holding fixture options as the moving part set-up. However, there are differences in the machine design, because a silicone roller is used in lieu of a flat hot-stamp die. The rubber, which is usually about 12.7 mm (½ in.) thick and bonded to a machined aluminum core, is available in many combinations of material, hardness, and length. Furthermore, after molding, each roller is precision ground to a concentric diameter in one of two configurations (Fig. 5.26).

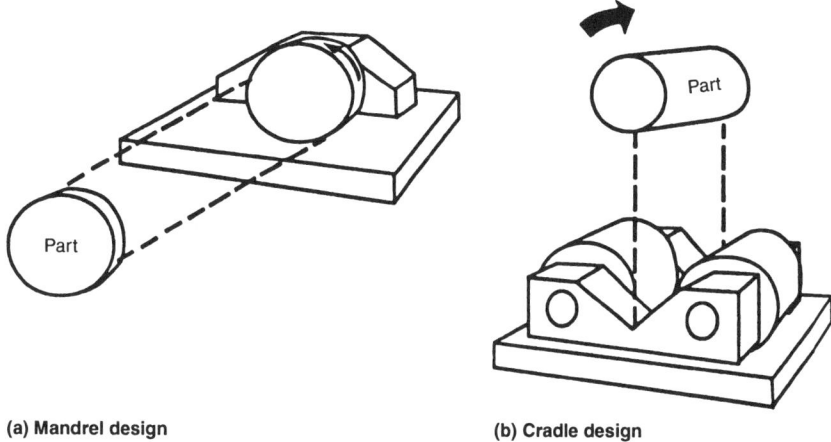

(a) Mandrel design **(b) Cradle design**

Fig. 5.25 Common part holding fixture designs. Courtesy of United Silicone, Inc.

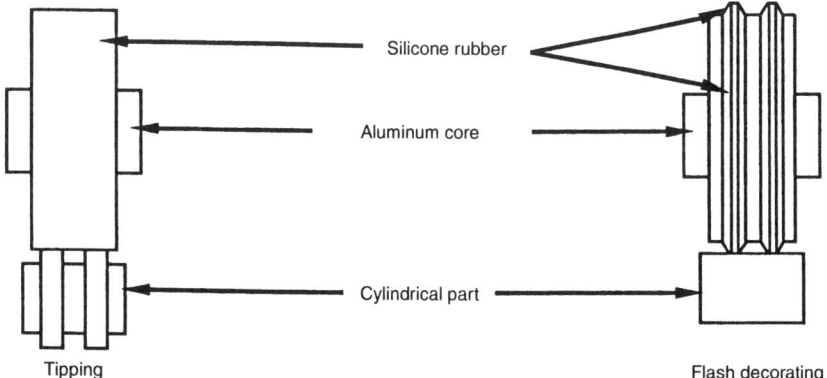

Fig. 5.26 Typical peripheral marking roller configurations. Courtesy of United Silicone, Inc.

Because of the extreme rubber thickness, which is required to help absorb the large amount of force exerted over a very small area, rollers are usually externally heated. The head of the machine includes a foil feed and a shroud with convection heat source. Constant roller rotation for even heat distribution is accomplished with an electric motor.

The term "stationary part" peripheral marking refers only to the actual application of hot-stamp foil. There is nothing stationary about the two common part transfer methods utilized to obtain improved production efficiency: a rotary index table and an over/under type indexing chain conveyor. Both methods allow for part loading and unloading during decoration when the fixture has stopped advancing, and this step can be done manually or with automatic feed and ejection devices.

Roll-on Decorating Technology

Where peripheral marking is concerned with cylindrical parts, roll-on decorating is ideal for applying folds to large area part surfaces, configured flat or with a slight contour. With this method, a silicone rubber roller applies heat and pressure to release the print medium onto the substrate. The advantage of this process is that the rubber roller material maintains line contact and pushes out trapped air between the printed medium and decorating surface so that air bubbles are eliminated.

Common Roll-on Decorating Methods. Just like peripheral marking, there are two distinct methods used when employing the roll-on technology, and both methods involve a heated silicone rubber roller. With the moving part method, the piece to be decorated is transferred past the roller while the foil is pulled through. With the moving head version, the part and foil remain stationary while the head travels in a box motion. A good example of a familiar application is a plastic television cabinet. Decorative foils, such as wood grain designs, can be applied to both sides and the top using either roll-on method (Fig. 5.27).

For both roll-on configurations, the head of the machine has a similar design. The roller is externally heated with a shrouded convection heat source, and even distribution is obtained through an electric motor-driven roller rotation. However, the similarities end there. The moving part method incorporates three different part transfer techniques. A special tensioning device is usually added to the foil feed to make sure the printed medium, as it is fed by the roller-to-part pinching action, is applied to the large area wrinkle free.

With the reciprocating table part transfer technique, one or more part holding fixtures are mounted to a platform that is typically driven hydraulically or with an electric motor coupled with a rack-and-pinion. An electric motor-driven belt conveyor and a roller-to-roller machine, used for decorating continuous pieces of extrusion, incorporate an uninterrupted part feed that facilitates efficient production output.

On the other hand, with the moving head roll-on method, the part holding fixture is mounted to a stationary table, or multiple fixtures can be incorporated with a rotary index table to increase productivity. Lateral roller head movement is commonly achieved using a ball screw drive with variable speed electric motor, while vertical action is accomplished with an air cylinder. This configuration usually includes a frame-mounted heat transfer indexer to accurately advance preprinted decals, and the indexer can be switched to time hot-stamp foil advance.

Part-holding fixtures are usually used in only two of the four roll-on decorating machine configurations discussed: reciprocating table and moving head. The key elements are much the same as those presented in the section "Vertical Stamping Technology" with two exceptions. Lead-on and lead-off supports help define the plane at which the roller will travel. Thus, roller damage and rollover at the edge of the part is minimized when the heated silicone first touches the piece and when it leaves after the decoration is complete. The second variation is a filler piece that is placed in any size part opening to support the foil. Without this flexibility, the foil carrier can become distorted from the heat, which can lead to wrinkles and then a poor finish (Fig. 5.28).

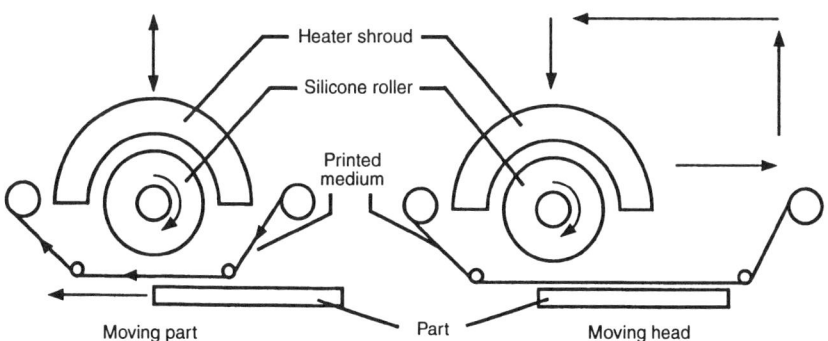

Fig. 5.27 Common roll-on decorating methods. Courtesy of United Silicone, Inc.

138 / Decoration and Assembly of Plastics

Regardless of the roll-on machine configuration, silicone rubber rollers are used to apply the decorative medium. The designs are consistent with those described in the section "Peripheral Marking Technology," except that contour decorating applications, where the silicone is precision ground to match the part shape, are more frequent (Fig. 5.29).

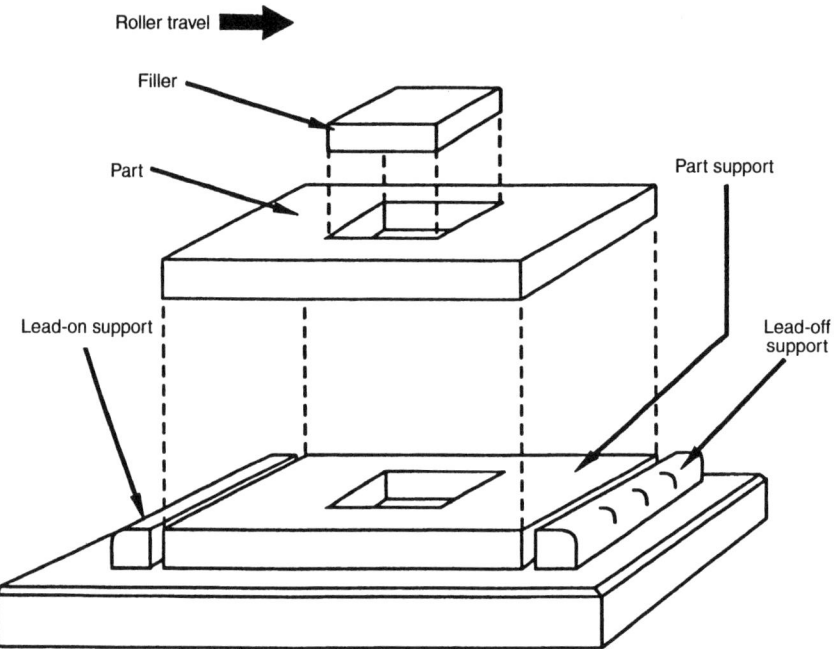

Fig. 5.28 Part holding fixture with lead-on/lead-off and filler. Courtesy of United Silicone, Inc.

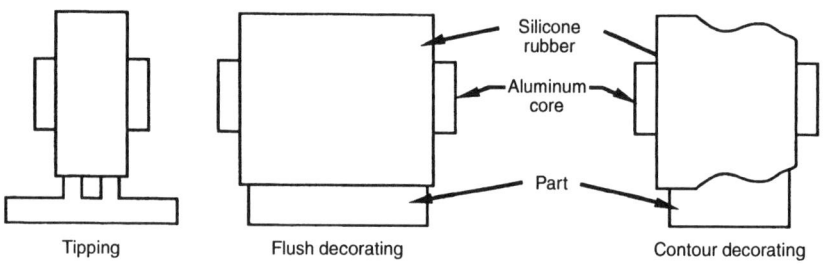

Fig. 5.29 Typical roll-on decorating roller configurations. Courtesy of United Silicone, Inc.

Although metallic and single color foils are used for roll-on applications, the most popular are multicolor designs, such as wood grains. Many shades of brown are layered with black to create the wood pattern and subsequently a heavy foil coating. This extra weight requires a thicker film carrier than is used for vertical or peripheral marking jobs. Similarly, the thicker carrier is also advantageous for other pigment or metallic foil colors, because it helps minimize wrinkling (Table 5.8).

Table 5.8 Hot stamping troubleshooting

Cause	Remedy
Incomplete hot stamped image	
Uneven die-to-part contact	Reposition fixture so decorating surface is parallel to a flat die or conforms to a contoured die
Plastic part contamination	Discontinue use of flow agents, anti-stat solutions, and/or silicone-based mold releases
Foreign particles on decorating part surface	Clean surface with white cotton cloth or glove
Insufficient fixture support permits the part to flex under the force exerted by the machine	Redesign fixture to provide rigid support under decorating surface of part and/or ensure that mandrel-type designs do not deflect
Air entrapment between foil and part surface	Redesign die face to include convex crown
Roll-over in the hot stamped image	
Uneven die-to-part contact	Reposition fixture so decorating surface is parallel to a flat die or conforms to a contoured die
Excessive force exerted by the machine	Decrease stroke length and/or reduce machine output
Die face is too hot	Reduce thermostat setting on the machine
Lengthy die-to-part contact duration (dwell time)	Reduce dwell time setting on the machine
Sinks in the decorating surface	Introduce makeready beneath the part in the areas that are hitting light, then reduce force as indicated above
Variations in the wall thickness from part-to-part	Switch to a dual durometer silicone rubber die
Edges of hot stamped image are not "sharp"	
Insufficient force exerted by the machine	Provide adequate stroke length and/or increase machine output
Plastic part contamination	Discontinue use of flow agents and/or silicone-based mold releases
Die face temperature is not hot enough	Increase thermostat setting on the machine
Short die-to-part contact duration (dwell time)	Increase dwell time setting on the machine
Foil stripping conditions	Slow head retraction and/or introduce head-up delay
Foil coatings are affected by lengthy heat exposure prior to the cycle	Use before/after foil selector to advance foil just prior to head descension
Poor foil-to-part adhesion	
Insufficient heat at the die face	Increase thermostat setting on the machine and/or move the thermocouple closer to the die face
Short die-to-part contact duration (dwell time)	Increase dwell time setting on the machine
Plastic part contamination	Discontinue use of flow agents, anti-stat solutions, and/or silicone-based mold releases
Foreign particles on decorating part surface	Clean surface with white cotton cloth or glove
Foil-to-part incompatibility	Switch to different foil formulation

Courtesy of United Silicone, Inc.

ACKNOWLEDGMENTS

The author wished to acknowledge the contributions of Mr. Eric Miles and United Silicone, Inc. (Lancaster, NY) to this chapter.

6

Pad Printing

For years, hot stamping and screen printing have dominated the decorating industry. Now, pad transfer printing is finding increasing favor for industrial and commercial applications because of its ability to print ink on a range of part surface geometries, from flat to complex shapes. Furthermore, new technology is available to make pad transfer printing more hassle-free by eliminating the need to constantly check ink viscosity levels.

Advantages of Pad Transfer Printing

The versatility of pad transfer printing is the most important asset. Other important advantages of pad transfer printing include:

- *Variety of part geometries:* Unique shapes as opposed to flat surfaces can be printed.
- *Variety of part surfaces:* Various degrees of textured surfaces can be printed.
- *Variety of materials:* Besides plastic materials, where the emphasis is on thermoplastics, thermosets can also be pad printed; in addition, glass, metal, wood, leather, and paper are decorated with success.
- *Duplication of fine detail:* Graphics with stroke widths as small as 0.076 mm (0.003 in.) can be accurately and consistently reproduced (Fig. 6.1).
- *Wet-on-wet capability:* Unlike screen printing, a second color can be printed over a first color immediately.

- *Four-color process:* The four primary colors (magenta, yellow, cyan, and black) can be printed in combination to reproduce images in all colors of the spectrum.
- *Permanent decoration:* Pad transfer printing ink has excellent adhesion to a variety of substrates and abrasion resistance.

Of course, the result of a system that is set up and maintained properly is a high-quality permanent decoration. Typical applications include symbols on automotive control panel buttons, characters on plastic keyboards for computers, and graduation markings on medical components.

Fig. 6.1 Pad printed and hot stamped parts. Courtesy of United Silicone, Inc.

Pad Printing Process

Step 1: Ink Flooding. The image to be transferred to the substrate is etched into a steel or nylon plate, commonly referred to as the "cliché" (Fig. 6.2). The entire top surface of the cliché, which is mounted in the inkwell, is flooded with ink by the flood bar. A stainless steel doctor blade then wipes the excess ink from the cliché surface, leaving ink only in the etched areas. After the cliché is wiped by the doctor blade, the surface of the ink in the engraving, which is exposed to the air, becomes more viscous and tacky as the solvents evaporate. As this occurs, the ability of the ink to adhere to the silicone transfer pad is improved.

Step 2: Pad Wetting. The pad is positioned directly over the cliché, pressed to pick up the ink, and then lifted away (Fig. 6.3). The physical changes in the ink that take place during flooding (and wiping), combined with the high surface tension of the silicone pad, account for the ability of the ink to leave the recessed engraving in favor of the pad.

Step 3: Head Stroke. After the pad has lifted away from the cliché to its complete vertical height, there is a delay before the ink is deposited on the substrate (Fig. 6.4). During this stage, the ink has just enough adhesion to stick to the pad (it can be easily wiped off, yet it does not drip). The ink on the pad surface once again undergoes physical changes: solvents evaporate from the outer ink layer that is exposed to the atmosphere, making it more tacky and viscous. Solvents on the inner surface migrate toward the pad, reducing the pad/ink adhesion.

Step 4: Ink Deposition. The pad is pressed down onto the substrate, conforming to its shape and depositing the ink in the desired location (Fig. 6.5). Even though it compresses considerably during this step, the contoured pad

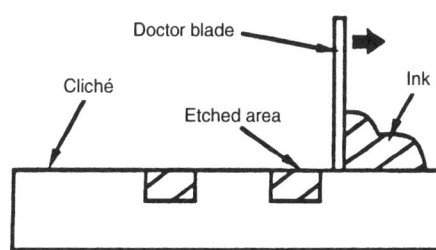

Fig. 6.2 Ink flooding

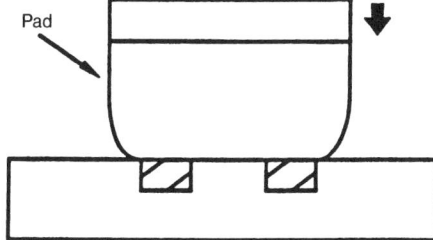

Fig. 6.3 Pad wetting

144 / Decoration and Assembly of Plastics

is designed to roll away from the substrate surface rather than press against it flatly. A properly designed pad, in fact, will never form a 0° contact angle with the substrate. Such a situation would trap air between the pad and the part, which would prevent ink transfer.

Step 5: Pad Release. The pad lifts away from the substrate and assumes its original shape, leaving all of the ink on the substrate (Fig. 6.6). As explained in Step 3, the ink undergoes physical changes during the head stroke and loses its affinity for the pad. When the pad is pressed onto the substrate, the adhesion between the ink and the substrate is greater than the adhesion between the ink and pad. This change results in a virtually complete deposition of the ink, which leaves the pad clean and ready for the next print cycle.

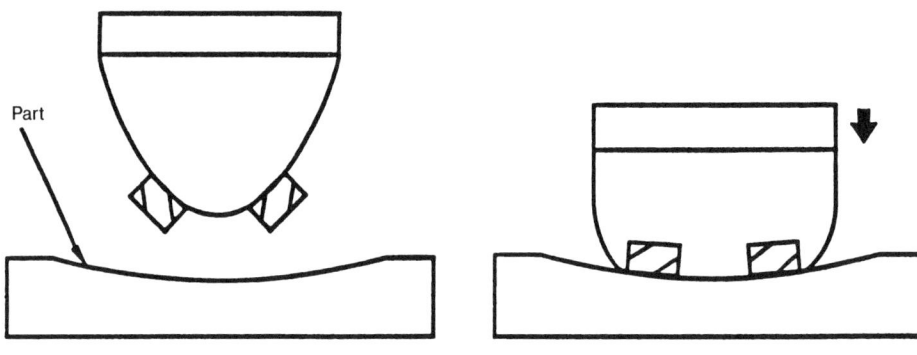

Fig. 6.4 Head stroke

Fig. 6.5 Ink deposition

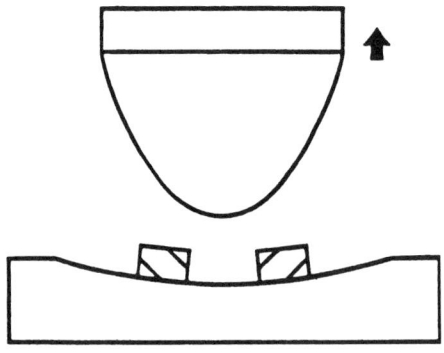

Fig. 6.6 Pad release

Table 6.1 provides suggestions for troubleshooting pad printing operations.

Vertical Printing Technology

Figure 6.7 compares three pad printing methods. Vertical-acting applications, similar to the one described in the section "Pad Printing Process," are by far the most common. However, for extremely high-speed production requirements, rotary methods can be utilized (Fig. 6.8).

Vertical machines are ideal for printing ink on a range of part surface geometries, from flat to complex shapes, and to a maximum of 90° of the

Table 6.1 Pad printing troubleshooting

Problem	Cause	Remedy
Voids and/or variable opacity in printed image with no ink retained on the pad surface	Pad is not broken in, and ideal amount of silicone oil is not present on the pad surface.	Compress the pad approximately 50 times to extract oil.
	Ink is too "dry" (thick) and is remaining in engraved cavity on cliché.	Add thinner to the ink a little at a time.
	Ink is displaced from the engraved cavity of cliché due to excessive doctor blade wiping pressure.	Reduce doctor blade wiping pressure.
	Ink is displaced from the engraved cavity of cliché due to shallow depth.	Engrave image cavity of the cliché to a greater depth.
Void in printed image with some ink retained on the pad surface	Ink is too "wet" (thin).	Use pad delay and/or add more ink to inkwell.
	Pad failure, there are no silicone oils on the surface to aid the release of ink.	Replace the pad, and make sure compression is no more than $\frac{1}{3}$ of its total thickness.
"Pin hole" size voids in the printed image with no ink retained on the pad	Entrapped air during ink pick-up	Reposition pad point so contact with image area is eliminated, and/or choose pad with larger contact angle.
"Hairy" image	Static charge	Use static eliminator on pad or in ink.
	Ink is too "dry" (thick)	Add thinner to the ink a little at a time.
	Image cavity on engraved plate is too deep.	Engrave image cavity of the cliché to a shallower depth.
Blurry image	Ink is too "wet" (thin).	Use pad delay and/or add more ink to inkwell.
	Pad is compressed too much and/or undersized.	Reduce pad compression to no more than $\frac{1}{3}$ of total thickness, and/or utilize larger pad.
	Part moves between first and second print on a double hit application	Redesign the holding fixture to locate the part better
Distorted image	Pad is compressed too much and/or undersized	Reduce pad compression to no more than $\frac{1}{3}$ of total thickness, and/or utilized larger pad.
	Insufficient pad support in part holding fixture	Add pad support to fixture in part openings, and/or around part parameter if pad wraps over edge.

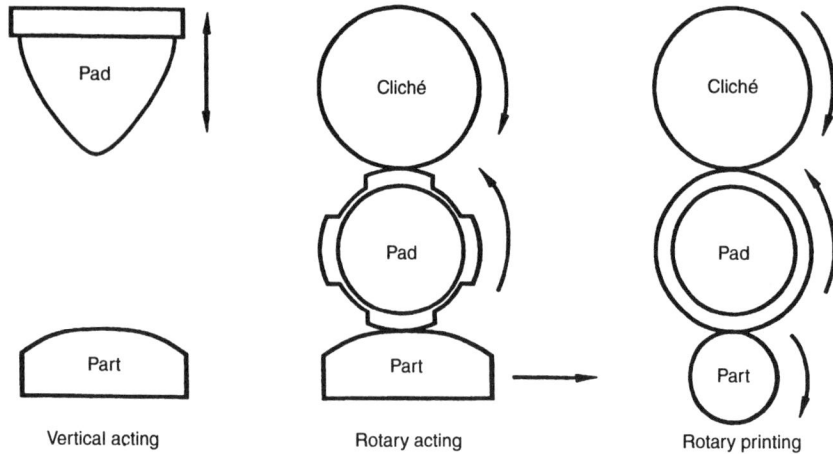

Fig. 6.7 Pad printing methods

Fig. 6.8 Sealed ink reservoir pad printing machine. Courtesy of United Silicone, Inc.

circumference of cylindrical or spherical part surfaces. Typically, the cliché remains stationary while the silicone transfer pad travels in an inverted "U" shape. The lateral and vertical pad movement is usually achieved with an electropneumatic or electromechanical drive (Fig. 6.9).

When considering equipment size, the most critical element is the amount of force that the machine can deliver. There must be a sufficient amount of force to easily compress the silicone transfer pad to cover the entire engraved image. Accordingly, equipment is constructed at various output levels to meet the range of requirements.

In many pad transfer printing applications, a single operator manually loads and unloads parts to a stationary fixture. Increased production output and convenience can be achieved by adding part transfer devices that permit loading parts while other parts are being printed. Common auxiliaries include rotary index tables or indexing conveyors. Labor content can be significantly reduced with the addition of automatic feeding and discharge equipment. Vibratory bowl feeders and vibratory stacker feeders are common, while pneumatically activated units and pick-and-place mechanisms are the most popular for part removal.

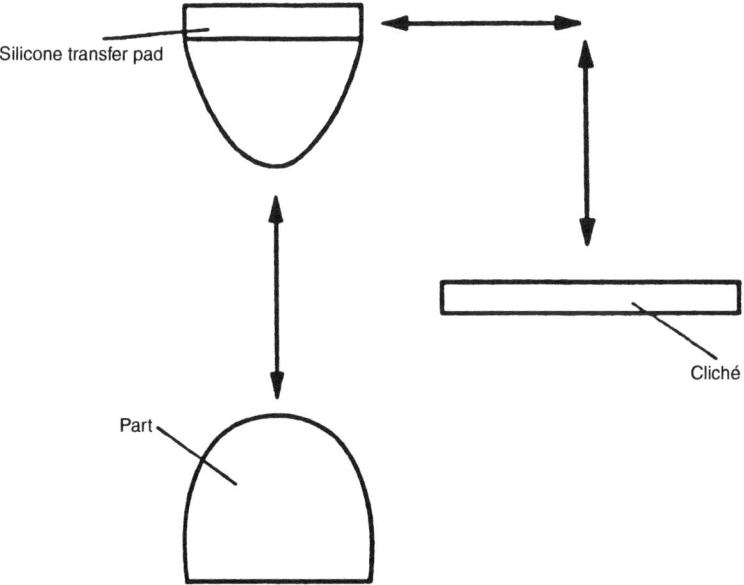

Fig. 6.9 Vertical pad transfer printing machine showing directional movement of the silicone transfer pad and cliché

148 / Decoration and Assembly of Plastics

Engraved Plate Technology

Perhaps the most important variable in the pad transfer printing process is the engraved plate, or cliché (Fig. 6.10). Typically, these are constructed in one of two different materials. Steel plates are extremely durable and ideal for large volume production runs. Nylon plates are popular for shorter runs, and they provide the opportunity for in-house photoengraving, which results in maximum flexibility.

Regardless of the material used, the resultant print will only be as good as the artwork. Without touch-up, the bottom line is that the cliché and the print will match the art exactly. Therefore, careful examination of the art, under a magnifying glass if possible, is recommended.

Once the artwork is finalized, some specialty image treatments should be considered before the plate is engraved. For some pad transfer printing applications that include images with long straight lines and/or large area graphics, ink may be displaced from the etched cavity prior to printing by

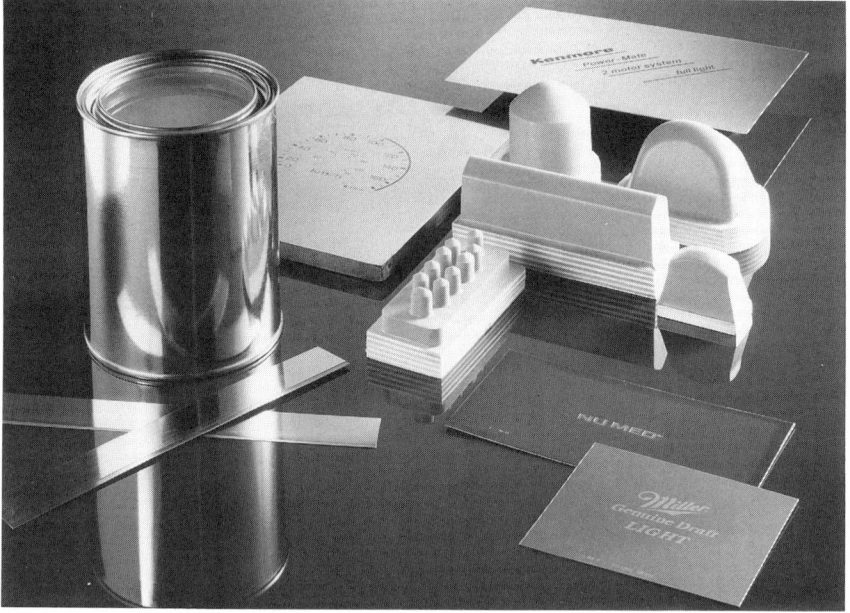

Fig. 6.10 Examples of engraved plates and other pad printing supplies. Courtesy of United Silicone, Inc.

the doctor blade or the silicone rubber transfer pad. The results are a reduction in the thickness of printed ink and poor opacity.

This problem may be solved by using one or both techniques shown in Fig. 6.11. The image can be oriented at an angle so that the doctor blade is supported properly as it passes. For large area graphics, the artwork can be "screened" using one of several dot patterns available. Then, when the plate is etched, the image is left with projections shaped like small, truncated cones. These projections help support the doctor blade and pad so that ink is not displaced from the etched area (Fig. 6.12).

The final determination for any cliché is the etch depth, which depends entirely on the image. The most common depths are shown in Table 6.2.

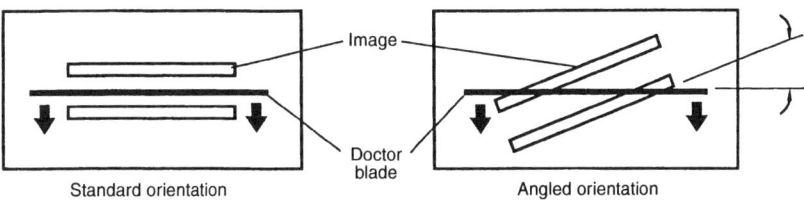

Fig. 6.11 Engraved plate technology angled image technique

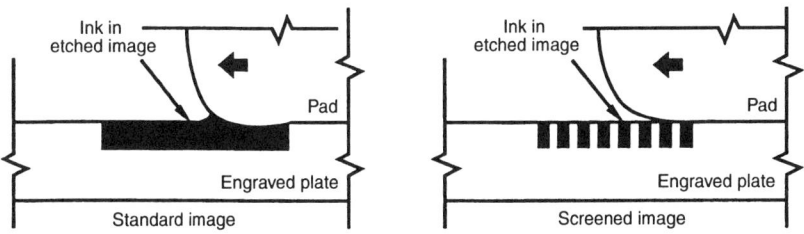

Fig. 6.12 Engraved plate technology screened image technique, usually used for graphics with long, straight lines

Table 6.2 Cliché etch depth and applications

Cliché etch depth, mm (in.)	Common applications	Character width, mm (in.)
0.0178 (0.0007)	Graphics with fine detail	0.076–0.793 (0.003 to $1/32$)
0.0254 (0.0010)	Graphics with normal detail	0.793–6.35 ($1/32$–$1/4$)
0.038–0.043 (0.0015–0.0017)	Graphics with large areas	>6.35 (>$1/4$ in.) and image is screened

Doctor Blade Technology

After the entire top surface of the engraved plate is flooded with ink, the doctor blade is the component that actually contacts the plate during the wiping stage to remove excess ink. The configuration of the stainless steel blade is critical to satisfactory doctoring, and different material thicknesses afford the luxury of matching the blade type to the application.

If the stroke width of any etched graphic exceeds 6.35 mm (¼ in.), then a thick material (about 0.508 mm, or 0.020 in.) should be used. With this configuration, the doctor blade is less likely to flex under the pressure of the wiping stroke and scoop ink from the engraved image. Conversely, a thinner blade (0.203 mm, or 0.008 in.) can be utilized for thinner images, but, regardless of material thickness, the wiping edge is chemically etched instead of ground, so that the contact surface is as smooth as possible (Fig. 6.13).

Doctor Blade Life. When installing a new doctor blade, the beveled wiping edge should be wiped with a Scotch-Brite (3M, St. Paul, MN) pad. This step cleans and polishes the contact surface of the blade to aid in satisfactory doctoring. The life of the blade will be increased if the doctor blade holder assembly is positioned so that the cliché is wiped with a minimum amount of pressure. The doctor blade should be replaced when ink streaking on the cliché cannot be eliminated with minor adjustments to the doctor blade holder. This situation is an indication of one of two things. Either the wiping edge of the doctor blade has been damaged, or the beveled area on the doctor blade has been almost completely worn away.

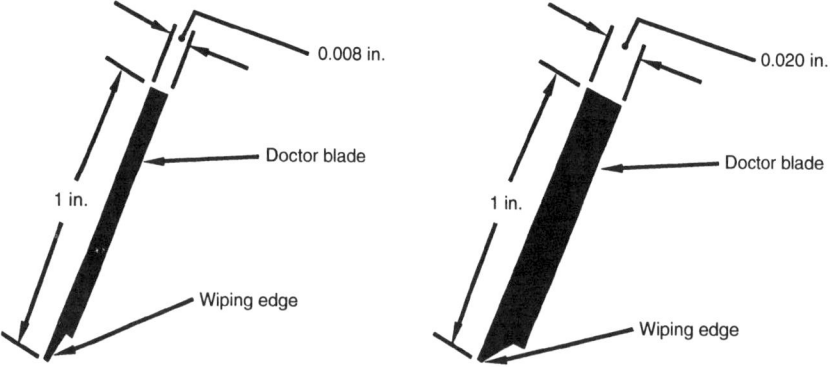

Fig. 6.13 Doctor blade configurations

Transfer Pad Technology

Silicone rubber transfer pads are available in an assortment of shapes and hardness. These choices are determined entirely by the characteristics of the decorating surface and the design to be printed. The pad acts as a transfer mechanism to move ink from the printing plate to the part being printed (Fig. 6.14). In order to do this, it should create a rolling effect when it is compressed onto the engraved plate, not a blotting action that can trap air. The face of the pad needs to be angled rather than flat, and some common shapes are shown in Fig. 6.15.

In addition to the overall shape of the transfer pad, the tip of the pad is also critical. A pointed tip is the most common, and the tip should not contact any part of the graphic when ink is picked up from the engraved plate. If it does, the pad will likely displace ink, which leads to a void in the printed image. For larger solid graphics, when pad tip contact with the image area cannot be avoided, a rounded shape should be used. The gentle contour results in less pad compression; therefore, the chance of ink displacement is minimized in the pick-up and printing positions.

Angles. The most common pad angles are 30, 45, and 60°. The smaller the image is, the higher the angle should be to allow more time for air to escape from the image as the pad is compressing and rolling. On the other hand, the more contour the part surface has, the smaller the angle should be so that less compression is required to completely cover the surface.

The pad thickness, or height, should be chosen so that during ink pick-up and printing the pad never compresses more than one-third of its total thickness. By following this rule, the bond life between the silicone rubber and the base will be maximized. Furthermore, over compression squeezes silicone oil from the pad, which will, over time, deter the proper release of the ink from the surface.

Sizes. When checking the printing size of the pad, it is important to note that a pad with more mass will tend to give less print distortion than a pad that is just big enough to fit the image. A smaller pad might not always cause print distortion, but a larger pad would generally be the best choice.

Hardness. Available in various levels of hardness, silicone rubber is able to conform to a range of part shapes, including severe contours. Picking the right silicone pad hardness depends on the shape and/or texture of the part being decorated and also the image being printed. There are four different hardness options ranging from 65 to 22 durometer on the Shore 00 scale. In general, a harder pad will deliver a "sharper" image, and a softer pad will

152 / Decoration and Assembly of Plastics

wrap around a contoured part better to allow for greater coverage. For reference purposes, 60 durometer on the Shore 00 scale is 12 durometer on the Shore A scale.

Fig. 6.14 Close up of a sealed ink reservoir. Courtesy of United Silicone, Inc.

Pad base materials for bonding silicone rubber pads commonly include wood, but aluminum is used for high tolerance mounting applications. Wood bases are constructed from a very good grade of 15.87 mm (⅝ in.) thick marine plywood, and a metal threaded insert is pressed into the center of the back of the base for mounting purposes. Aluminum should be used when the location of the drilled and tapped mounting hole in the base is extremely critical. Otherwise, wood is the material of choice, because the silicone bonds better to it. Wood is more porous than aluminum, and it will not oxidize.

Note: To help maximize the life of a pad, it should be stored in an enclosed cabinet to minimize exposure to dust and other potential contaminants. The ideal storage temperature is 10 to 21 °C (50 to 70 °F), and the pad should sit on its base to deter the migration of silicone oil to the surface.

Ink Technology

In order to achieve optimal printing results, it is essential to use special pad printing inks. These inks are aromatic, and they cure by evaporation. They feature extremely high pigmentation levels, because only a slight amount of ink is transferred during the process. Virtually any Pantone (Pantone, Inc., Carlstadt, NJ) color number can be formulated, and metallic inks

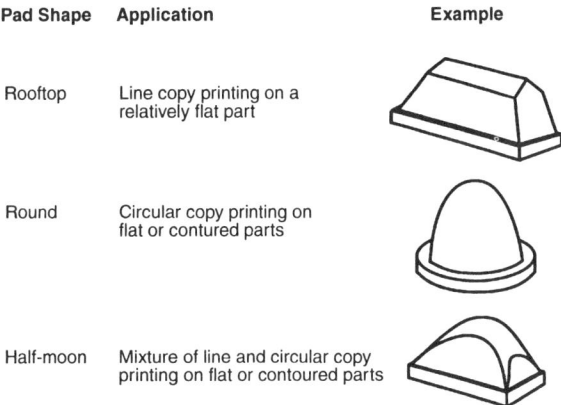

Pad Shape	Application	Example
Rooftop	Line copy printing on a relatively flat part	
Round	Circular copy printing on flat or contoured parts	
Half-moon	Mixture of line and circular copy printing on flat or contoured parts	

Fig. 6.15 Three common transfer pad shapes and applications

can also be specified. Pad transfer printing inks are made from different additives and the following:

- *Binder:* Made from resins and dissolved in solvents
- *Pigment:* Responsible for the color shade of an ink
- *Filler:* Promote certain properties, such as opacity, viscosity, and abrasion resistance

In addition, an assortment of auxiliaries, mixed in the proper amount with the ink, provide for easy ink processing:

- *Thinner:* Used strictly as a means transporting the ink from the engraved plate to the part by changing the viscosity
- *Retarder:* Used to slow the evaporation of the thinner. Rarely used, but extremely beneficial on humid days
- *Hardener:* Used as a catalyst to help harden the outer skin of ink to promote improved abrasion resistance

By selecting from these components, the ink will be tailored to a specific application and will only adhere to a specific family of substrates.

Pad transfer printing inks are categorized by the number of components they contain. A single-component ink is adjusted, prior to printing, to the proper viscosity by adding thinner. For enhanced abrasion resistance, hardener is mixed with thinner and ink in a two-component format.

The consistency of pad transfer printing ink constantly changes. After mixing, single-component inks have considerable longevity as long as they are maintained properly, either manually or automatically. On the other hand, two-component inks have a "pot life" and will become too viscous within 8 h. After printing, all ink types dry to the touch quickly, and drying time can even be accelerated by directing heated forced air across the graphics. Finally, when an ink fully cures, all solvents have evaporated, optimal adhesion to the substrate has been obtained, and physical testing can be conducted.

When the printing substrate is known, then an ink designed to adhere to the material can be selected. Sometimes there is more than one ink type for a specific substrate. Also, it is important to remember that thermoset materials and substrates from the polyolefin family (polyethylene and polypropylene) must usually be pretreated to obtain good adhesion.

Ink suppliers are now challenged by environmental regulations, which stipulate that the heavy-metal content of pigments must be below specified levels. The changeover, however, is much more complex than just substituting some organic pigments for the heavy-metal variety. For example, the organic types have lower ultraviolet resistance than their counterparts, so a stronger inhibitor has to be added. Also, the two systems reflect light differently, which can result in color shade and/or opacity variances.

Part-Holding Fixture Technology

The various components of a fixture should be constructed directly from the parts, not from a part print, and three key elements should be considered (Fig. 6.16–6.18). First, the decorating surface should be presented as parallel as possible to the cliché surface. Next, the fixture should hold the part firmly in place throughout the print cycle, and it should be designed so that the part-to-part location is consistent. Third, support behind the decorating surface is not mandatory for smaller image applications as long as the part is rigid enough. However, for bolder/large graphics or textured surfaces, where extra pad pressure is required, the part should be supported.

Fixturing components are produced in a variety of materials including aluminum, steel, PVC, nylon, Teflon (E.I. Du Pont de Nemours & Co., Inc.,

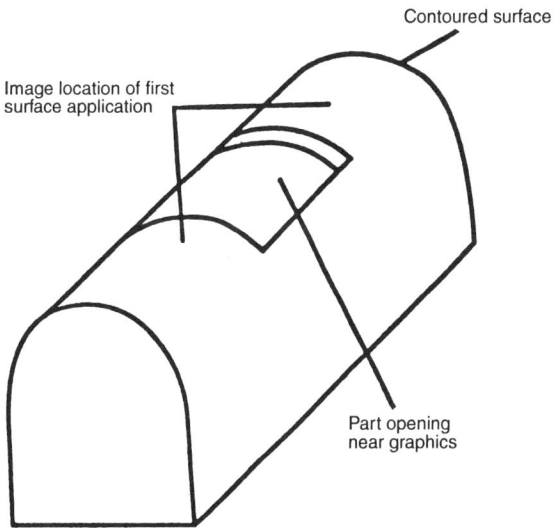

Fig. 6.16 Evaluating the part

156 / Decoration and Assembly of Plastics

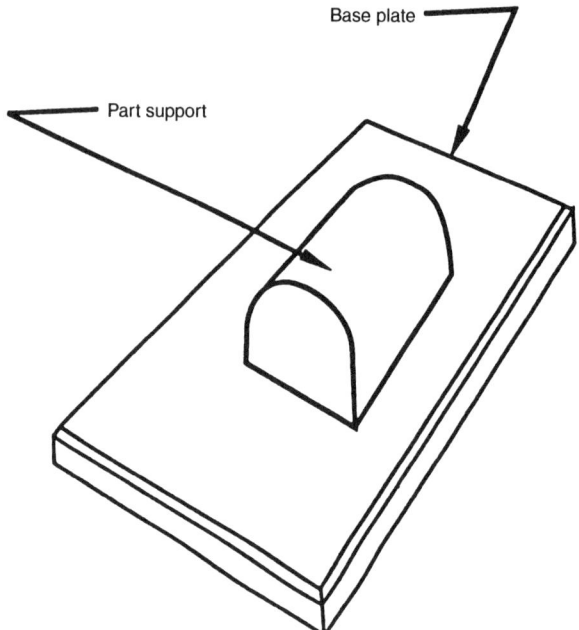

Fig. 6.17 Part-holding fixture

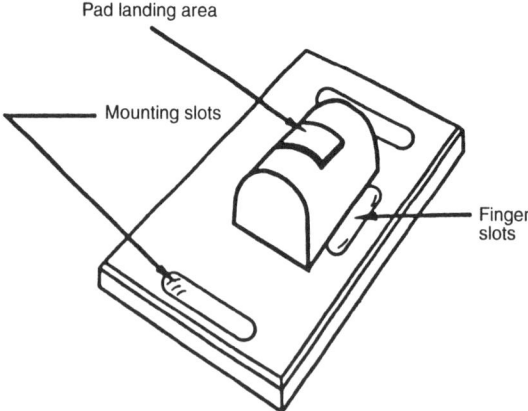

Fig. 6.18 Part-holding fixture options

Wilmington, DE), and a range of rigid and resilient casting compounds. For part-holding fixture base plates, aluminum is selected over steel, because it is lighter, durable, easy to machine, and will not rust. Similarly, aluminum is also popular for part supports. However, PVC is recommended when clear or high gloss substrates are decorated, because it will not scuff plastics as long as it is kept clean. Nylon and Teflon are like PVC, but they are also ideal for automatic part discharge applications because of a low coefficient of friction. Finally, for many contoured applications that require internal support, a coating of urethane is cast to the shape of the part and bonded to a metal base. The advantage of this is that a tighter fit is provided than any other material that is machined to the shape.

Finally, when a graphic is near the edge of a part, a "landing" area for the pad should be constructed in the fixture. The landing should fill an opening in the part and/or surround the perimeter of the piece. It should be situated at the same level of the part surface to support the pad so that the silicone does not stretch over an edge and/or into an opening. As a result, print distortion, caused by stretching is eliminated, and pad cuts can be avoided if sharp edge contact is avoided.

New Viscosity Control Technologies

To solve problems often associated with pad transfer printing operations, two new approaches are available to control ink viscosity. The sealed ink reservoir and ink monitoring pump offer innovative technologies that eliminate the need for periodic ink maintenance and permit continuous operation for long durations. Additional benefits include reduced printing rejects, minimized downtime, improved quality and consistency, and increased overall production efficiency. However, it is important to point out that these advantages are only obtained with single-component ink systems. If hardener is added to any formulation, the viscosity will exceed practical levels within hours, despite outside efforts to control.

Sealed Ink Reservoir. Unlike the open inkwell configuration, where the ink is continuously exposed to the atmosphere and constant thinner evaporation results, the sealed ink reservoir contains the ink in a vessel that can be compared to an inverted paint can (Fig. 6.19). The reservoir, and specifically its steel doctoring ring mounted around the open edge, maintains complete contact with the engraved plate throughout the flooding and wiping stages of a machine cycle. This sealing action virtually eliminates solvent evaporation

158 / Decoration and Assembly of Plastics

and associated fumes, and once thinner is initially mixed with the ink, subsequent additions of thinner may not be required.

Ink Monitoring Pump. Used in conjunction with any pad transfer printing machine that incorporates an open inkwell, the ink monitoring pump automatically maintains the viscosity of an ink at a specified level, which permits continuous printing operation for long durations (Fig. 6.20). A typical unit features an ink monitoring vessel, with a motor-driven agitator to con-

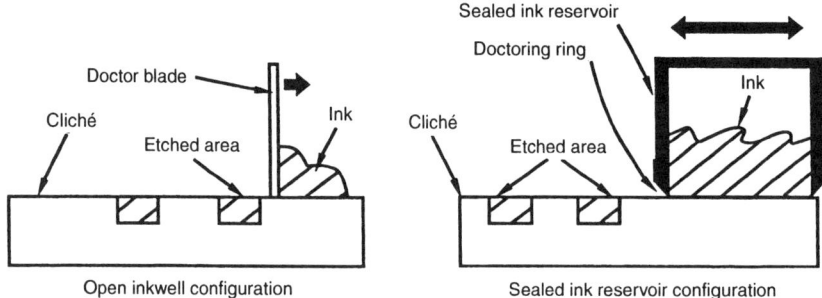

Fig. 6.19 Open inkwell and sealed ink reservoir configurations

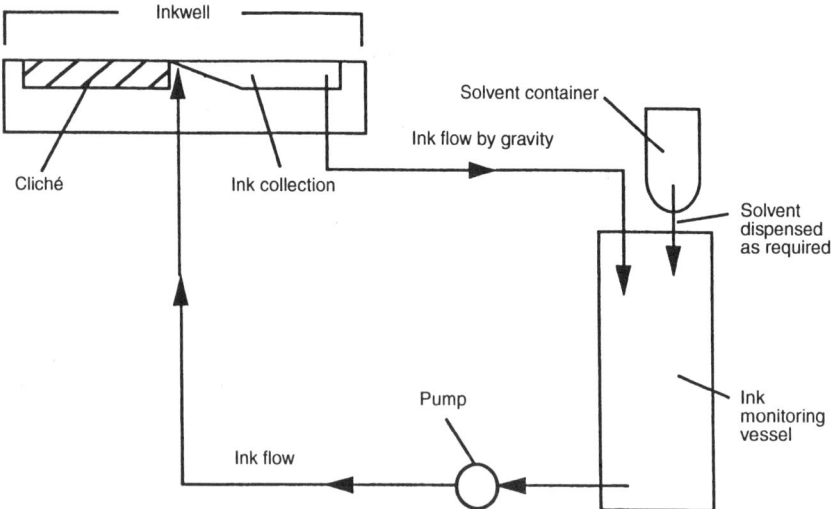

Fig. 6.20 Ink monitoring pump flow

tinuously mix a sizeable quantity of ink and monitor the viscosity. As required, solvent is dispensed into the ink to maintain a preset level, and then the ink is pumped to the inkwell before returning to the unit by gravity.

ACKNOWLEDGMENTS

The author would like to thank Mr. Eric Miles and United Silicone, Inc. (Lancaster, NY) for contributions to this chapter.

7

Metallization of Plastics

Plastics are nonconductors of electricity, which allows them to be used in many unique applications, such as electronic devices, protective housings, and advanced military systems (stealth vehicles). There are many applications that demand plastic materials to be electrically conductive or highly reflective. Compact discs, automotive lighting systems, and chrome-like decorative surfaces for auto and truck grills have unique demands for plastic materials (Fig. 7.1), and all of these applications require a metal layer to be somehow applied to the surface of the plastic. The use of metallized plastics is categorized as follows:

- *Functional:* Used to reflect or conduct energy, such as light or electricity, in electromagnetic interference (EMI) reduction and radio frequency interference (RFI) reduction
- *Decorative:* Used to create an appearance, such as auto/truck grills
- *Both:* Provides both a functional and decorative service

Although there are many ways to achieve the functional and decorative metallization of plastics and plastic products, the main methods include hot stamping (see Chapter 5), printing and coating, vacuum metallizing, electroless plating, and electrolytic plating. The latter three metallization processes are discussed in this chapter.

Vacuum Metallization

Many parts used in visible applications (such as automobile ornamentation and trim) require some sort of shiny metallic appearance. The electroplating

162 / Decoration and Assembly of Plastics

process has historically applied this finish to most parts (whether metal or plastic). Vacuum metallizing was developed as an alternative method of coating plastics with a metallic finish. Since the mid 1950s, vacuum-metallized plastic has competed with plated plastics and castings for automotive interior applications. Recent metallizing developments have increased the durability of the coatings. Vacuum metallizing has been extended to exterior automotive applications, such as grill units and light bezels. Most vacuum metallizing research and development has been accomplished by coating suppliers and vacuum equipment manufacturers to meet automotive requirements.

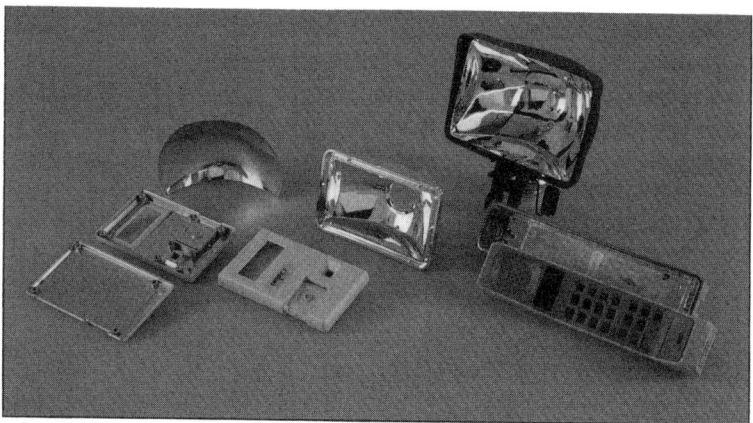

Fig. 7.1 Vacuum-metallized products. Source: Vacuum Platers, Inc., Mauston, WI

Vacuum metallizing was thought to be a process limited to coating relatively inexpensive plastic parts. Many of the technological advances developed for plastics can be utilized on metals. Application of metallizing to metal substrates, such as steel and aluminum, has received limited attention. The emphasis in metallizing has been to simulate chromium for automotive use. However, gold, copper, silver, brass, and a wide variety of colored metallic finishes can be duplicated for furniture, hardware, appliances, and other industries where decoration is necessary.

Virtually all plastic substrates can be metallized, except polyethylene and urethanes; however, the most popular substrates are styrene, acrylonitrile, acrylonitrile butadiene styrene (ABS), acrylic, nylon, polypropylene, polycarbonate, Noryl (General Electric Co., Pittsfield, MA), and phenolics (Fig. 7.1).

Vacuum metallizing is a physical, rather than electrochemical, process of depositing a metallic film onto a prepared surface. Deposition takes place within a high vacuum. Normally, the substrate is prepared with a lacquer to form a base for the metallic film. The parts are then put into a vacuum chamber, where metal is evaporated and condensed onto the base coat (Fig. 7.2–7.4). A nonmetallic top coat and, in some instances, overlay coats are applied to finish the process. The metallized layer is thus sandwiched, depending on the application.

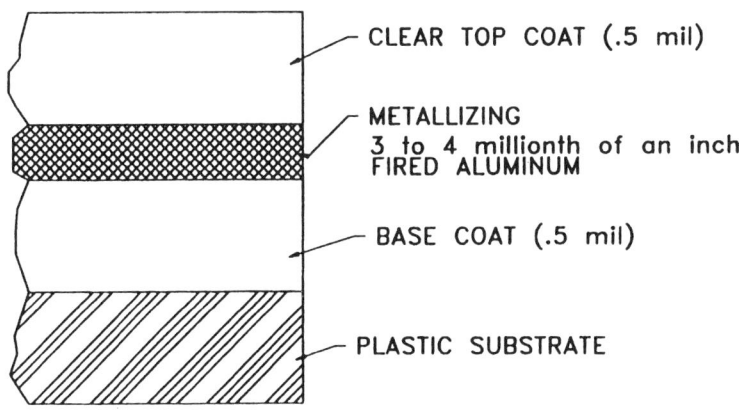

Fig. 7.2 Vacuum-metallized plastic

164 / Decoration and Assembly of Plastics

Vacuum metallizing is essentially a batch-type operation. However, equipment systems have been developed to eliminate handling between loading of racks, base coating, baking, metallizing, top coating, overlaying, and unloading. Coating systems vary considerably according to specific application.

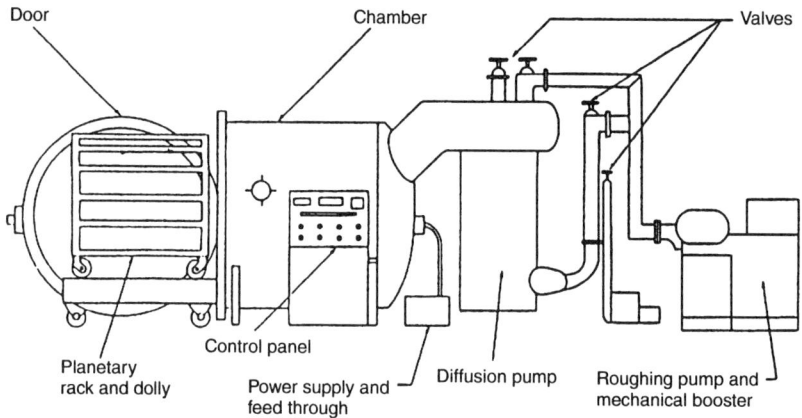

Fig. 7.3 Common 183 cm (72 in.) horizontal metallizing chamber. Source: Vacuum Platers, Inc., Mauston, WI

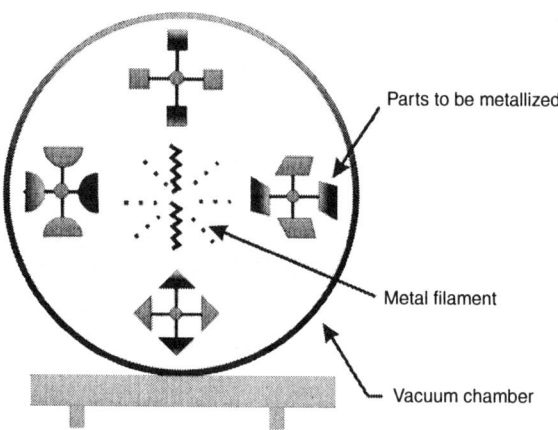

Fig. 7.4 Vacuum metallizing chamber

Vacuum Metallization Process

The major steps in vacuum metallizing are cleaning, racking, chamber evacuation, and vaporization.

Cleaning. All parts must be clean in order to assure adhesion and eliminate coating defects, such as "fisheyes," dust, and inclusions. They should be completely free of mold release agents, gross soils, oxides, and other reactive films. Castings, steel stampings, and spinnings should be smooth, clean, and free of porosity. They usually do not need buffing or polishing, because surface defects normally are covered and filled by the base coat. Phosphate and chromate dips are recommended to assist adhesion of the base coat.

Plastic components should normally be molded without mold release agents. There are paintable-plateable mold release sprays, which can sometimes be used. However, release agents that contain silicones should absolutely never be used.

Racking. Parts have to be held securely, because they go through various coating and metallizing processes. Usually, spring clips are mounted on individual rods or reels of rods, which are compatible with the processing equipment. Part design is important, because clip marks or scars are left where the part is attached to the rack. The attachment points should not be visible during service. Close attention should be paid to critical tolerance surfaces. Parts should not be affixed to the rack in these areas, because buildup is hard to control. Various masking devices, plugs, and caps may be used to shield critical areas.

Base coating is a painting operation, as is top coating. Base coat lacquers can be applied by all conventional methods, and parts may be flow coated, sprayed, or dipped. Paint equipment manufacturers have developed various types of special automated equipment specifically for the metallizing industry, which mate paint application systems with the vacuum systems. The coating must be fully cured, using either batch or conveyorized ovens. Curing times normally run from 30 min to 2 h, depending on the type of resin coating and curing temperature.

Chamber Evacuation. After the base coat is fully cured, the racks of castings are loaded onto carriages (of either stationary or planetary design), which hold the racks inside the vacuum chamber. Most chambers are cylindrical in shape and are mounted horizontally for easy insertion of the carriage through the entry door. Chambers vary in size from small, laboratory bell jars to large, conveyorized continuous units with load docks and chambers with varying degrees of vacuum in series. The most common model is a

180 cm (72 in.) inside diameter (ID) horizontal batch-type chamber, with lengths ranging from 110 to 180 cm (42 to 72 in.). After the carriage with part racks is inserted, the door is shut, and the pumping cycle begins.

Pumps. Three different types of pumps are normally used to achieve the high vacuum necessary for metallizing. The three pumps are arranged in series and are connected by valves. The first stage of the operation involves the use of a mechanical rotary piston pump along with a roots-type booster pump. About 99% of the air in the chamber is removed, lowering the pressure to 100 µm of mercury, which is the rough pumping stage. One or more high-capacity diffusion pumps are then used to reduce the pressure to 0.5 µm. This stage is called the diffusion or foreline stage, and it is the point at which the vaporization cycle occurs. Pumping times can be from 10 to 30 min, depending on system capabilities, system condition, and atmospheric conditions. In some installations, turbomolecular and cryogenic pumps have replaced diffusion pumps and reduced pumping times.

Vaporization. Any metal can be vaporized; therefore, any metal, including alloys, can be vacuum metallized. Aluminum is the most commonly used metal in decorative applications because of its physical properties, economic considerations, and required equipment.

Bus bars with tungsten wires or filaments run down the center of the chamber. Aluminum canes, clips, or ribbons are attached to the filaments. The carriage drive rotates the parts around the diameter of the chamber. When the electrical current to the filament is turned on, resistance heats the filaments. When the filament temperature reaches 660 °C (1220 °F), the aluminum melts and flows out (wets) along the filament. The aluminum vaporizes (flashes) off the filament when the filament temperature reaches 1200 to 1500 °C. The aluminum vapors leave the filaments in straight lines and hit the parts being rotated on the carriage. The vapors condense on the parts, because the parts are much cooler than the filaments.

Planetary carriages are used when it is necessary to expose all surfaces of parts to the aluminum vapors. Contours and recessed areas are covered, because the coating is still fluid when it condenses and tends to form a uniform layer on the part. The actual flashing takes about 10 s, while heat-up and prewet cycles take about 20 s.

The vacuum is broken by shutting pump valves and letting air into the chamber through another valve until normal atmospheric pressure is attained. The carriage with the now metallized parts is removed from the chamber, because the parts are ready for top coating.

Controls for the evacuation and vaporization sequences may be manual, semi-automatic, or completely automated. Automatic sequencing eliminates the cost of an operator and possible human error in the various procedures.

Resistance firing is not the only possible method of vaporization. Sputtering, electron-beam vaporization, ion plating, high-frequency resistance, radiant heating, and induction heating are other possible methods. These are "exotic" and are not discussed here.

Top Coating and Overlay. An organic top coat is used to protect the thin layer of metal. A proper top coat is essential to obtain the brilliantly polished aluminum appearance desired. The various aspects of top coating can be a science unto themselves. In addition to simulating all the bright metallic finishes, satin or subdued effects can be achieved by using semigloss rather than clear top coats. The top coat can be either clear or tinted to appear to be copper, brass, gold, or colored metallic finishes.

Tinted coats should always be spray coated, unless a special effect is desired. The higher specific gravity of the pigment causes it to accumulate in recesses when flow coated. Areas where the coat is thicker, such as edges, are also darker due to particle attraction. Flow-coated top coats must be clear, and the dip should be dyed in a resin-die solution to attain the proper tint and film color uniformity. Flow coating is the most economical system with regard to material and use, and it is often the best system for covering complex shapes. Spraying is usually best for parts that have large flat surface areas.

Overlay coatings are used to increase the film thickness, add to the durability, and increase abrasion resistance. Ultraviolet (UV) stabilizers may be added to improve the endurance of the film under direct sunlight. Clear two-component top coat systems using epoxy or urethanes are used for additional durability and are required for automotive exterior applications.

Plasma polymerization is another top coating process that applies the top coat to the parts immediately after metallizing and while the parts are still in the vacuum chamber. The UV curable coatings are also gaining acceptance in the metallizing industry because of their added durability and energy savings in application.

Decorative overlays can be sprayed or silk-screened over metallized components to achieve additional effects. Automotive and appliance trim is often screened with decorative trim or operating instructions.

Zinc Die Castings. Applying conventional metallizing to castings has not yet received the attention that steel and plastic parts have. Thin-wall zinc die castings are a natural for vacuum metallizing. Two cost-influencing variables of metallizing are weight and surface area, which can be processed through the system. Prior to thin-wall technology, the weight-to-surface-area ratio of zinc die castings was relatively high compared to that of plastics. Weight was the limiting factor, because most of the production equipment had been designed for plastics. Batch weights being equal, per chamber load, many more plastic parts could be processed. Because the thin-wall parts are much lighter, many more castings can be processed per load, reducing costs.

Weight is not necessarily the limiting factor. With denser castings, all load-bearing parts of the metallizing system should be beefed-up to withstand the additional weight. The racks, holding clips, coating equipment drives, carriages, and chamber equipment should be evaluated for the ability to withstand the fatigue of the additional weight.

In most instances, a buffing or polishing operation can be avoided with zinc die castings. The base coat buildup has a leveling effect and covers minor surface imperfections. If paint overlays are required, adhesion problems are eliminated because of the resin-to-resin bond with the top coat.

If it is necessary to recycle reject finished castings, any commercially available zinc-safe paint stripper will remove the coating. The castings can then be recycled through the metallizing process with virtually no castings lost because of finish defects.

The melting point of zinc allows it to withstand higher curing temperatures than plastics without losing its dimensional stability. As a result, tougher thermosetting coatings can be used for base and top coats. Lower coefficients of expansion and contraction allow vacuum-metallized zinc to better withstand thermal stresses. Water absorption by an electroplated or vacuum-metallized plastic part can cause swelling that will cause the coating to microcrack or blister, which eventually will lead to loss of adhesion and peeling. Automotive headlight and tail-lamp housings are ideal applications involving thin-wall zinc and vacuum metallizing.

Vacuum metallizing, like die castings, is a high volume process, and systems can be engineered to be compatible. A standard 180 cm (72 in.) ID batch-type chamber is capable of processing over 30,000 110 g (4 oz) castings per 8 h shift, which corresponds to a yearly production capacity of 7,200,000 pieces.

Tests

Vacuum-metallized steel and plastics have met tough nonvisual test requirements.

Tape adhesion required a minimum of 95% retention of coatings when the tape was applied to a crosshatch surface.

Humidity was tested with 96 h exposure at 38 °C (100 °F) and 100% relative humidity with no loss of adhesion, blistering, or other appearance changes.

Gravelometer, for abrasive resistance, had parts stand up to a minimum rating of 5 when tested under SAE 1400 at –18 °C (0 °F).

Thermal cycle exposure at 77 °C (170 °F) for 20 h was followed by a hold at room temperature, 20 h at –29 °C (–20 °F), and 4 h at room temperature. The cycle was repeated three times, and the parts showed no evidence of loss of adhesion, cracking, or other appearance changes.

Cleanability included the top coats resisting attacks from a wide variety of cleaners and solvents.

Wear was tested, and all coatings exceeded requirements for resistance to abrasion of the top coat and metal film.

Salt spray tests resulted in no blistering or corrosion in scribed areas and no loss of adhesion after 336 h of neutral salt spray. For reference purposes, lines are scratched through the coating down to the substrate in cross-hatch pattern.

Copper-accelerated acetic acid-salt spray (CASS) tests showed no blistering or corrosion in scribed areas and no loss of adhesion after 48 h.

Weatherometer. Exposure to UV was maintained for up to 1600 h without evidence of deterioration of top coat or change in appearance.

A major telephone manufacturer investigated the feasibility of replacing copper, nickel, and gold electroplating (overlaid with clear urethane coating) on decorator series zinc telephone components. Vacuum metallizing met or exceeded the following:

- Initial appearance standards described in ASTM B 456
- Salt spray test (CASS) for 4 h according to ASTM B 368
- Cyclic temperature test as in ASTM B 553
- Adhesive tape tests (minimum 95% retention) as in ASTM special publication 500
- Pencil hardness test (No. 2 hardness minimum) as in ASTM special publication 500
- Water immersion as in ASTM D 870
- UV exposure (100 h, no fading; 234 h, slight fade on high edges; 500 h, covered with a glass panel) as in ASTM D 795
- 30,000 phone hang-ups passed with coating abrading the acrylic handle

The variety and flexibility of vacuum coating systems can be altered to obtain the specifications desired. Results of more in-depth testing and coating applications should indicate that actual performance standards will be better for vacuum-metallized zinc than plastics in all areas.

Costs

Since the mid 1950s, vacuum metallizing has gradually become accepted as the standard finish, replacing electroplating. The automotive, toy, lamp and lighting, trophy and award, novelty, giftware, hardware, and other highly competitive industries have made the switch due to cost pressures.

Cost differences vary according to size, complexity of part, and specific application. However, generally speaking, vacuum metallizing has cost advantages over electroplating. Chemicals, environmental clean-up, capital investment, and tooling all cost more with electroplating. Direct labor, handling between the various coating processes, is greater with vacuum metallizing. This factor is the primary reason why continuous and semicontinuous hands-off systems have been developed.

There are no simple guidelines for engineers to follow when specifying what materials, processes, and finishing methods are to be used to manufacture parts. In-depth studies must be made weighing all elements and advantages of various alternatives. For this reason, product manufacturers should consult the providers of finishing services while still in the early stages of product design. Metallizers, platers, anodizers, and others have extensive experience that can impact product appearance, function, wear life, and cost reduction.

Vacuum metallizing technology has increased rapidly with tougher organic coatings, more automated equipment, and better metallic evaporants. While successful in some specific applications, it still has a way to go before equaling electroplating on all applications. But as time passes, more traditionally electroplated components are being replaced with vacuum-metallized parts. Indeed, vacuum metallizing has a very "bright" future.

Electroless and Electrolytic Plating of Plastics

Product designers have had a love-hate relationship with chrome-plated products. In the 1950s and 1960s, automobile manufacturers embraced large chrome-plated steel bumpers, grills, and trim. Plastic usage in the automotive industry accelerated in the 1960s, and the need to chrome-plate plastics was developed. The quality of plated plastics was poor and contributed to the cheap image suffered by the plastics industry when used as a substitute material.

Improvements in plastic materials and the plating processes contributed to a second surge of plastic usage in the automobile industry in the 1970s and 1980s. It was, however, the breakthrough in the quality and acceptance of the chrome-plated plastic grill that marked the turning point in metallization of plastics. By the late 1980s, virtually all auto and truck grills were manufactured using plated plastics. The acceptance of the lightweight/robust product was considered by the prestige auto manufacturers mark of engineering achievement and quality improvement.

Selecting a Plastic Material

Plastics materials that are selected for products to be electroplated must possess a key attribute not required in general material selection scenarios: the plastic material *must be etchable*. The word "etchable" means that the plastic material can selectively have one or some component of the material removed at the surface to be plated.

The most popular plastic used in the plastics plating industry is ABS or variants of ABS. ABS is called a terpolymer, because it is comprised of three different polymer components. The "A" represents acrylonitrile, which is a key component in clear plastics, such as acrylics (Plexiglas [AtoHaas Americas, Inc., Philadelphia, PA], Lucite [E.I. Du Pont de Nemours & Co., Inc., Wilmington, DE], Acrylite [CYRO Industries, Rockaway, NJ]). These materials are known for both clarity and outdoor weatherability. The "B" represents butadiene, which is a rubber component that affords ABS good impact strength. Additionally, butadiene is the component that can be chemically etched or removed from the surface of an ABS part. The butadiene can be removed without significantly altering the properties of the remaining ABS or the plastic part. The "S" represents styrene, which is also known for its clarity. Styrene has a low impact strength and low cost relative to the acrylonitrile and butadiene.

Together the ABS components exhibit great teamwork referred to as "synergism." Synergistic behavior is when the final product/material properties exceed the sum of the individual properties (i.e., 1 + 1 = 3). The proportions of ABS material components can be varied to provide different properties. A simple graphical explanation may be considered:

- ABS: A lower cost ABS due to greater styrene component

- ABS: A higher impact strength ABS with possible improved etching qualities due to the higher butadiene component
- ABS: An improved ABS in terms of weatherability and colorability due to the higher acrylonitrile component

Preplate

There are two key preplate steps in the electrolytic plating process for plastic materials: etching and electroless plating.

Etching is a surface preparation process that takes advantage of the specific plastic to be plated. In the specific instance of ABS, it is the butadiene that is etched or removed from the surface of the plastic part. The selective removal of the butadiene renders the surface of the plastic part with a pockmarked appearance. These pockmarks, pockets, or undercuts (Fig. 7.5) act as mechanical footings for all the deposited coatings that are subsequently applied.

The etching process, like all of the plating process steps, is accomplished in chemical tanks called baths. The etch bath usually consists of a concentrated acid solution, most commonly chromic or sulfuric acid. The variables, in addition to the acid type and concentration, are time in the bath and bath temperature. Proper control of these variables will result in control of the amount of butadiene etched, which, in turn, defines the size of the pockets. The size and shape of the pockets will determine how well the subsequent coatings adhere to the plastic.

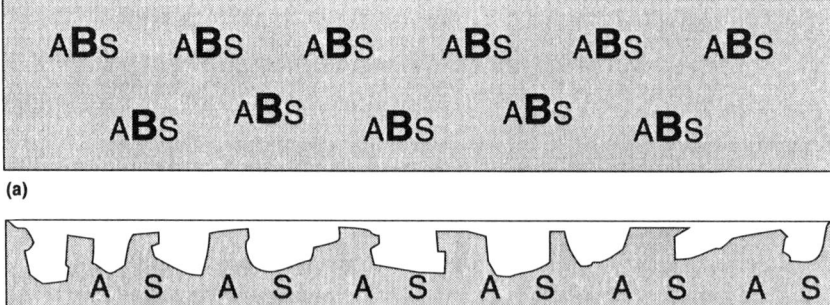

Fig. 7.5 ABS surface (a) before and (b) after etching

Electroless plating is the second step in the preplate process. Electroless, as the name implies, does not use electricity. The electroless process uses chemical reactions to deposit and initial layer or layers of metal onto the plastic (Fig. 7.6, 7.7). The electroless deposition layers are held onto the plastic part by the etched undercuts. Once the plastic part is properly electroless plated, it is rendered conductive and can then be subsequently plated electrolytically.

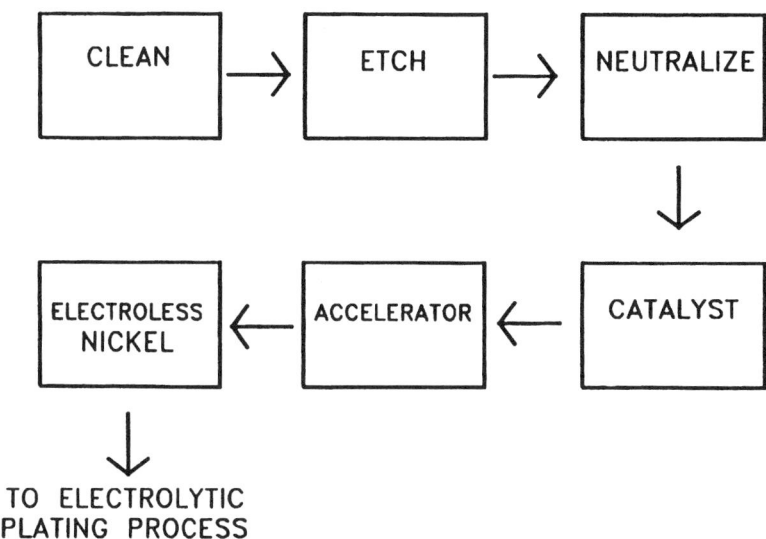

Fig. 7.6 Flow diagram of electroless plating

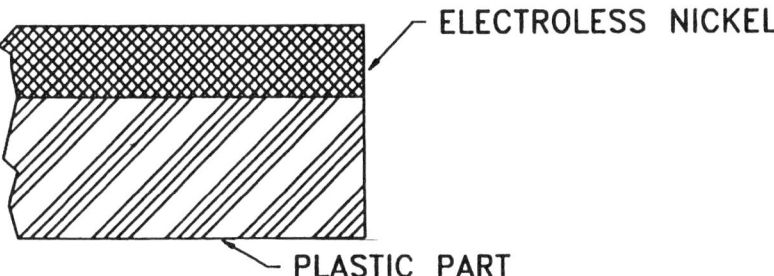

Fig. 7.7 Electroless plating

The first step in the electroless plating process is to place the etched plastic part into a bath of catalyst. The catalyst is a palladium material (palladium chloride) in the form of suspended microscopic particles (called a colloidal suspension). These particles find their way into the etched undercuts now present on the surface of the plastic (Fig. 7.8). The second step is to have the part placed in an accelerator bath. The accelerator is a chemical bath that activates the palladium-based material now nestled in the undercuts of the plastic part. Once activated, the palladium can act as a nucleation site. In other words, the palladium will now be able to beckon the metals out of solution (nickel and/or copper) that are suspended in subsequent baths (Fig. 7.9). This process of chemical deposition of metals is referred to as an "autocatalytic" process. It will continue unless controlled by the time and temperature of the various baths.

After the plastic part has been activated (i.e., had its initial metal coating(s) precipitate on its surface), it is now conductive. The electroless metallic layers are not robust. It is the subsequent layer of metal coatings that will provide the bright durable coating.

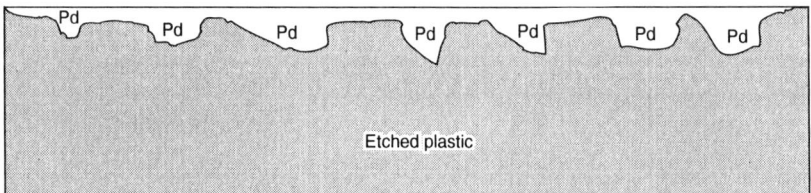

Fig. 7.8 Palladium on the surface of the etched plastic part

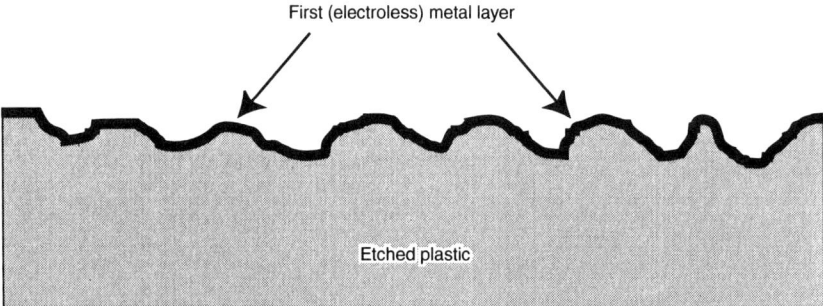

Fig. 7.9 First (electroless) metal layer

Electrolytic Plating

The electrolytic plating of a plastic part utilizes electrical current to precipitate additional metals out of a solution and layer them onto the electroless metallized surface of the plastic part (Fig. 7.10, 7.11). The electrolytic coating is applied in several layers by moving the plastic part through various chemical baths. Although the sequence and nature of these baths may vary between various plastic platers and processors, there are some basic baths used by most plastic plating operations.

Copper strike is the first electrolytically deposited layer of copper placed over the electroless nickel/copper layer. The previously electroless metallized plastic can now be grounded (given a negative charge), and the copper metal can be made an anode (given a positive charge). The tank solution has to be capable of carrying a current, which is accomplished by adding salts (copper sulfate) and acids (sulfuric acid) to create electrolytes (Fig. 7.12).

Acid Copper Bath. In this bath, the plastic part receives its first decorative metallic layer. All the prior metallic layers will appear dull and unattractive. The copper, deposited in the acid copper bath, has a bright appearance and a metallic luster. The appearance of this layer will have a direct affect on the appearance of all subsequent layers added to the plastic part. The electrolytes in this bath are carefully monitored, and additional organic brighteners are added to help control appearance and quality.

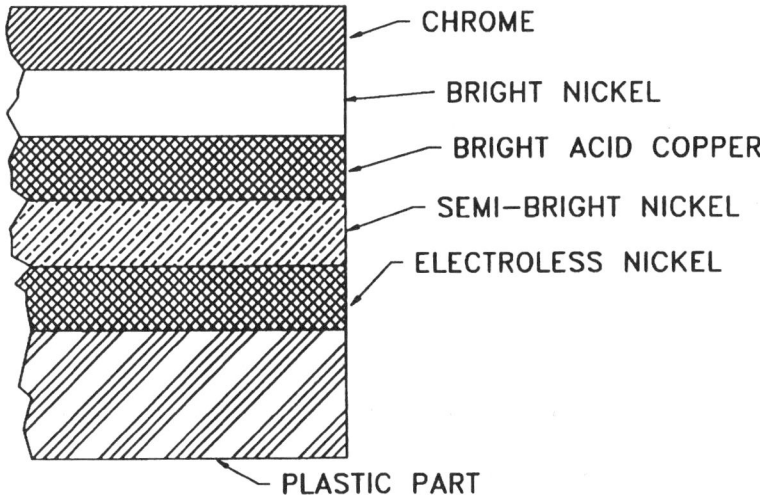

Fig. 7.10 Electrolytic plating

```
FROM ELECTROLESS
    PROCESS
       │
       ▼
┌──────────┐     ┌──────────────┐     ┌──────────┐
│ SULFURIC │     │ SEMI-BRIGHT  │     │ SULFURIC │
│   ACID   │ ──▶ │    NICKEL    │ ──▶ │   ACID   │
│   ETCH   │     │              │     │   BATH   │
└──────────┘     └──────────────┘     └──────────┘
                                            │
                                            ▼
┌──────────┐     ┌──────────────┐     ┌──────────┐
│          │     │    BRIGHT    │     │  BRIGHT  │
│  CHROME  │ ◀── │    NICKEL    │ ◀── │   ACID   │
│          │     │              │     │  COPPER  │
└──────────┘     └──────────────┘     └──────────┘
```

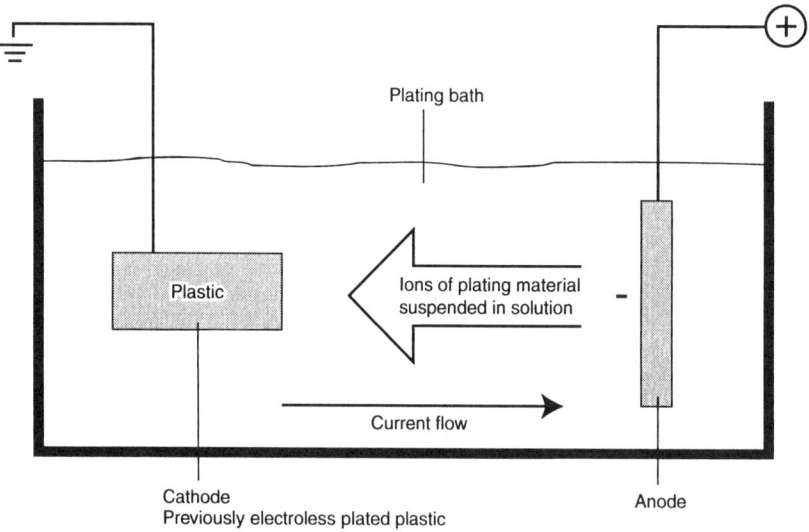

Fig. 7.12 Electrolytic plating action

A semibright nickel bath utilizes different electrolytes (nickel sulfate and nickel chloride). The nickel, deposited on the previously applied layers, will be the first bright and shiny metallic layer. It is this appearance that yields the semibright characterization of the nickel layer.

The bright nickel bath provides the plastic part with its bright decorative finish. For some plating operations, the bright nickel is the last metallic layer to be applied. The chemistry of the baths is similar to the semibright nickel baths, with the exception that additional brighteners are added to the tank. Most plated plastic parts will continue to have other metallic layers deposited beyond the bright nickel bath.

Microporous Nickel-Dur-Ni. This is a microfine-discontinuous nickel bath used predominately for exterior automotive applications. This layer is a nonconductive layer of microfine particles, and it prevents a broad corrosion of the plating by inhibiting corrosion growth.

Chrome is usually the final step in the plating process. The bath consists of chromic and sulfuric acid electrolytes. The anode in this bath is predominately lead with less than 10% tin. There may also be a catalyst to support the brightness of the plate.

Washing and neutralizing are present between most electrolytic baths to prevent cross contamination. After the final (chrome plating) bath, the plated plastic parts are dipped in several baths to rinse off any residue or electrolytes. Additionally these final baths are intended to remove any water spots.

Product designers tend to incorporate complex decorating systems for plastic parts. These systems often include painting on chrome-plated plastic. The final wash and neutralization baths ensure that any coatings applied to the chrome plating will be as durable as the chrome plating itself.

Racking Plastic Parts for Plating

The quality of the plating process depends on the design and proper use of part racks as much as it depends upon the quality of the various baths. The racks used to hold the plastic parts as they journey through the various plating tanks serve three key purposes:

- Holding the parts securely
- Providing the electrical connections to the plastic parts to be sure they are negatively charged (grounded)
- Designed to incorporate a wirelike anode to allow plating to occur on complex shaped parts

Figure 7.13 illustrates how a rack may be designed to provide these three functions.

Measuring Quality

The American Society of Electroplated Plastics (ASEP) has developed a series of conditional references that are helpful in assessing the end-use environment for plastic parts that are electroplated. There are four basic service conditions (SC).

SC 4 involves very severe service conditions where impact, abrasion, and scratching are very likely. Additionally, exposure to corrosive environments and temperature cycling may be present. Product examples include automotive exterior components, decorative plumbing, and marine applications.

SC 3 incorporates severe service conditions that may include exposure to outdoor weathering, detergents and cleaning solutions, other chemicals, and possible temperature cycling. Product examples include hospital products, exterior furnishings, toys, and cookware.

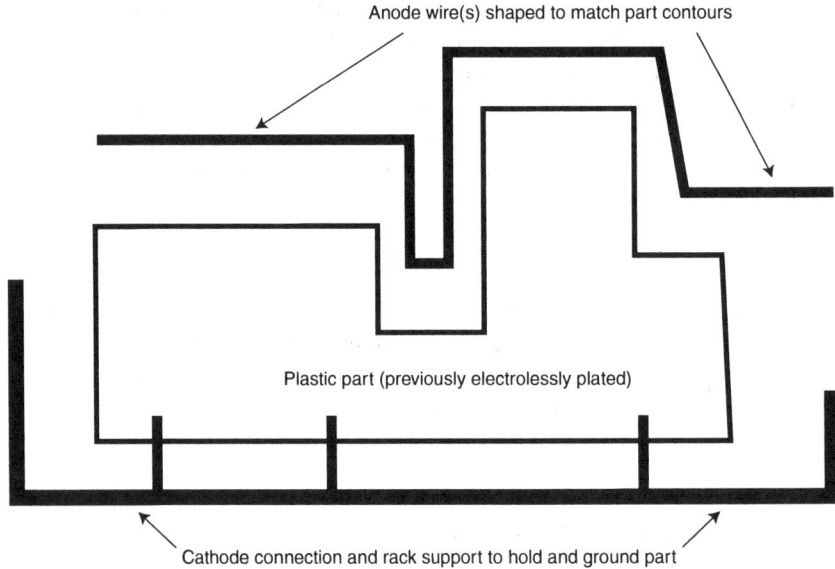

Fig. 7.13 Plating rack

SC 2 involves moderate service conditions that are mainly indoors. Some exposure to moisture and thermal cycling may occur. Product examples include kitchenware, cosmetic packaging, trim bezels, and control knobs.

SC 1 allows mild service conditions, which are indoors and have minimum wear, moisture, and temperature cycling. Product examples include decorative and novelty items.

Design Issues and Concerns

As the science of designing and manufacturing of plastic parts improves, engineers strive to develop design rules and guidelines to assist in the production of quality plastic products. When metallizing using the electrolytic process is concerned, these rules and conventions must be selectively modified if a quality plated plastic part is to be the result.

Figure 7.14 indicates a few design accommodations. Plastic part designs must incorporate subtle wall and feature (projections and holes) changes for the metal to be uniformly applied to the part surface. Abrupt design changes will result in a chrome "burr," an unsightly blister of metal that will diminish the plated products appearance and value. Additionally, it should be understood that plating does not cover-up or fix cosmetic problems that exist prior to plating. In fact, the opposite is true: surface anomalies will be exaggerated after plating.

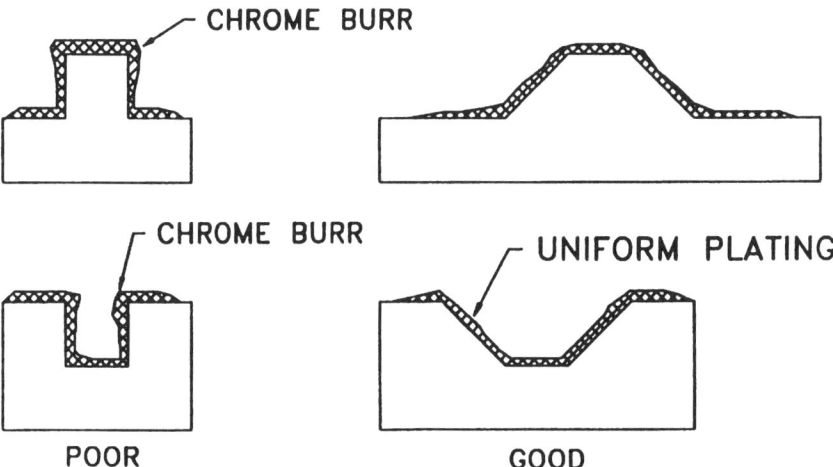

Fig. 7.14 Common plating problems when the transition to holes and projections is too great

Design should refrain from attempting to plate large areas of a plastic product. It is well known that the human eye is very perceptive to viewing flatness, and surface anomalies will be obvious. It is also known, however, that the human eye can be easily deceived by curved surfaces; therefore, large flat areas should be redesigned to incorporate a subtle crown or dome effect. This detail masks the appearance of any subtle surface defects.

8

Painting, Coating, and Printing

Plastic products often require some degree of coloration. Whether to create an aesthetic appearance, such as a body panel for an automobile, or simply to distinguish one part from another, there are several methods of applying color to plastic products.

Internal Coloring

When the polymer is manufactured, additives incorporated internally create the plastic compound (see Chapter 2). Often, one of these additives is a colorant. The type and amount of colorant depends on the plastic material. The objective is to provide a color to the plastic that exists throughout the plastic, which can be a truly beneficial attribute. An example is vinyl siding. The customer appreciates that the color of the siding exists throughout the siding so that any nicks or scratches will be less obvious.

The advantage of internal coloring must be factored by the fact that the polymer, being a color itself, may act as a filter. The resulting color of the plastic may be different from that of the colorant being used. Some polymers, such as acrylic, polystyrene, and polycarbonate, are clear and, as a result, minimally filter the color when internal colorants are incorporated. Other polymers, such as polyethylene and polypropylene, are translucent, and they tend to filter the internal colorant to some degree. This step makes internal coloring difficult to match, because there is a considerable amount of polymer filtering taking place. Finally, some polymers tend to be opaque, such as ABS, PPO, and glass-filled materials.

When coloring matching, especially to other materials, is required, painting is often the preferred method of adding color to a plastic part. Internally colored plastic products can be accomplished by one of several techniques.

Precolored Plastic

Plastics processors that do not want to compound and color raw plastic materials have the option of purchasing precolored plastic pellets from the resin/plastics supplier. These precolored materials are often much more expensive than basic uncolored plastic. The plastics part manufacturer also will have little or no latitude in controlling the color throughout the process.

In-House Coloring

If the cost of purchasing precolored plastic material is too expensive and/or there is a desire to have more control over the coloring process, plastics part manufacturers can explore in-house coloring. The cost savings of purchasing uncolored plastic is significant, and the manufacturers assess the requirements to color plastic as part of the plastic part molding process. If, for example, the manufacturer is purchasing polystyrene in precolored batches of red, yellow, and blue, the cost of purchasing the total amount of plastic in natural (uncolored) bulk quantities may easily justify the capital costs associated with the loading and mixing systems, as well as the expense associated with the labor to execute the coloring of the plastic. Hidden costs include the risk of incorrectly mixing or matching the color. The latter concern is usually addressed by purchasing a coloring matching booth or a hand-held colorimeter.

Colorants for plastics usually come in several forms.

Dyes are soluble colorants used to color some plastics after the part has been molded. Dyes are not used as internal colorant. The dye is used in much the same way as an article of clothing would be dyed. The soluble dye colorant is absorbed into the plastic part. The quality of coloring and color matching with dyes is poor.

Pigments are insoluble colorants added to or mixed with plastic pellets prior to molding. Because the pigment is insoluble, it may require additional additives to support proper dispersion through the plastic melt as the material is processed. The internal nature of these colorants is appealing to the processor from an ease of use consideration. Again, the resulting quality of the coloration is a function of the filtering of the plastic resin. Proper selection of colorants for plastic materials can result in high quality colors and color matching.

Painting

Although painting plastic parts appears to be a process that does not take into consideration all of the unique aspects of plastics (i.e., ability to be colored internally), many processors and designers consider it to be the only coloring process that can assure a proper color match between adjacent plastic and nonplastic parts located on an automobile or piece of furniture.

Successful painting on plastics requires an understanding of four key areas: the properties of the substrate (the plastic part), the properties of the paint or coating system, the application process for delivering the paint onto the plastic, and the end-use application and environment.

Understanding the Properties of the Substrate

Although a significant amount has been written about plastics and the surface phenomena associated with these unique materials (see Chapters 2 and 9), it is worthwhile to highlight a few key points:

- Plastics materials and surface conditions vary from material to material.
- Many plastics require surface activation or excitement to allow paints and coating to properly adhere.
- The surface of the plastic part to be painted must be clean and free of any process oils or lubricants.
- Internal additives may not manifest themselves as problems until months or years of use.
- Aging of the plastic substrate may render the bond between a plastic and the paint/coating weak.
- Molded-in stress within the plastic part may stimulate crazing and reduce the adhesion between the plastic substrate and the paint/coating.

Types of Paint

The majority of paints used for plastics (and most other substrates) are, in fact, polymers themselves. Thermoset and thermoplastic paints and coatings are used in very specific applications. The advancements in polymer chemistry that yield stronger plastic materials also yield more durable paint systems. A few of the main paint systems are described below.

Multicomponent. As the name implies, multicomponent paint systems have more than one component. Usually the bulk of the paint volume is the resin, and the minority component is a catalyst or hardener. The two-part paint systems are most often thermosetting polymers that require special delivery systems and possibly curing areas to affect the crosslinking of the polymer chains. The two-part paint systems are often more expensive than the single-component systems; however, they are more durable.

Epoxy paint systems, like adhesive counterparts, offer stronger adhesion, high temperature resistance, and good chemical resistance. Epoxy paints historically have been used in applications where the substrate is relatively solid or rigid, and the end-use application is in a volatile environment, such as a marine or industrial setting.

Polyurethanes have found a true market in automobile applications, where the substrate may be somewhat flexible, such as ground affect components and front and rear fascia (Fig. 8.1). Polyurethane paint systems are both robust and flexible and can be applied to a wide variety of substrates. Polyurethane usage will, no doubt, increase as costs decrease.

Fig. 8.1 Painted flexible fascia

Enamels historically referred to paint systems with a chemical reaction that results in a surface film that leaves a high-gloss surface when dry. While this description is still true of an enamel system, the word "enamel" is used more as an adjective to describe any high-gloss paint.

Waterborne or water-based paint systems, as the name implies, utilize water as the medium to suspend the paint particles. The use of water, as opposed to solvents, has been a direct result of environmental concerns. Many paint systems utilize solvents to suspend the paint particles (see the section "Solvent-Based"), which require significant capital investment to be in compliance with environmental codes.

Unfortunately water-based paints are less durable than the solvent-based systems, so there has been less waterborne replacement of solvent-based paint systems. Laboratory work on the waterborne systems will, no doubt, eventually develop more durable systems. Because plastic parts are less tolerant of solvents than other substrate materials, the plastics industry is very supportive of the conversion from solvent to water-based paint systems.

Acrylic paint systems, which gained popularity in the 1970s, are water soluble, easy to clean up, quick drying, and environmentally safe. There is minimum odor associated with the application. The nature of acrylic polymers (water clear) is an excellent medium for suspending paint particles and a resultant brilliant, rich color with depth. However, acrylic paint systems are not as robust as the solvent-based counterparts.

Latex is essentially rubber particles suspended in water. Homeowners know latex paint systems are a great alternative to oil or solvent-based paints. The water-based aspect of latex paint results in low/no odor, easy clean-up, and quick drying. Additionally, the latex rubber allows the paint to stretch, which makes it excellent for substrates that may tend to expand and contract with temperature changes.

Solvent-based systems are used in the majority of painting applications on metal and nonplastic substrate. The mechanics of solvent-based paint systems is that the paint particles are suspended in a compatible solvent. The paint is applied to the substrate, and the solvent supports a uniform coating. The solvent is allowed to flash-off (evaporate), and the paint particles fuse together. In some situations, the solvent will slide off the surface of the substrate to affect a better paint-to-substrate bond.

The chemical nature of the solvent and the fact that it becomes airborne when evaporating are major environmental concerns. Special air handling

186 / Decoration and Assembly of Plastics

equipment and operator safety gear is required to be in compliance with environmental codes.

Powder coats are fine particles of either thermosetting or thermoplastic materials. The powder coat is applied without the use of any solvents. In many applications, the fine particles are electrostatic charges, and the substrate to which they are applied is oppositely charged. The particles tenuously cling to the substrate that is placed in a heating apparatus that fuses the particles to form a uniform coating.

The benefits of powder coating are apparent. No solvent or water is used, and compliance to environmental codes is accomplished with less capital investment. However, there are a limited number of powder coat systems and substrate on which they can be used. Another disadvantage is that the overspray can result in a higher-cost coating system than more conventional painting systems. The furniture industry has embraced the powder coating technology more than any other industry segment.

Soft paints were designed to provide a softer feel (over hard plastic components) for the consumer. Additionally, the soft paint system feels warmer to the touch than unpainted plastic. The automobile and furniture industries have embraced soft paint systems for arm rests, dash components, and auto interior details (Fig. 8.2).

Soft paint systems are usually two-component polyurethane systems that blend flexibility and toughness, while providing a leatherlike appearance and feel. Clearly, replacing leather with a soft paint is a true cost savings.

The application thickness of a soft paint system is approximately 3 mils, which is significantly greater than conventional paint systems. The added

Soft touch painted surface

Fig. 8.2 Soft touch painted surface provides warm interior colors and feel to automobiles and trucks

thickness is required to achieve the full leatherlike effect desired by consumers.

Paint System Components

Paint systems, whether solvent-based or water-based, have a lot in common. Figure 8.3 shows the binder, the pigment, and voids. The binder, which is the matrix that holds the pigments together, is usually solvent- or water-based. The pigment is the color component of the paint and is dispersed evenly throughout the binder. Voids are the areas between the pigment. There is no coloration in the void, so the color density is inversely proportional to the volume of voids. More voids result in lower color density.

Figure 8.4 illustrates the effectiveness of a paint system by showing what may happen as a painted surface has light shown upon it:

- *Point A:* If light is reflected at the surface, then the surface is shiny.

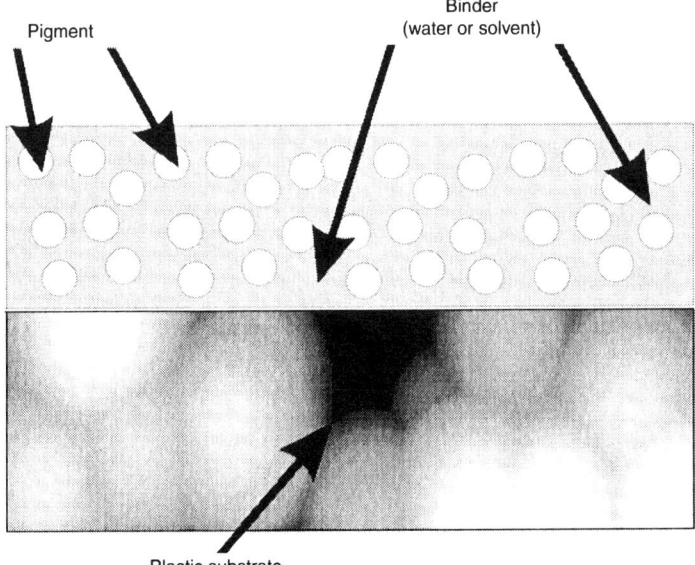

Fig. 8.3 Paint system

188 / Decoration and Assembly of Plastics

- *Point B:* If light is reflected by the pigment, then the surface is colorful (at least the color of the pigment).
- *Point C:* If light is absorbed by the pigment, then the surface has a dark appearance.
- *Point D:* If light is reflected by the surface of the substrate, then the "paint" was clear. Although this condition may result in the appearance of depth, there are no color attributes associated with the paint pigment.

Conformal Coatings

Conformal coatings are a category of coating systems that are used for functionality rather than decorative appearance. The largest market for conformal coating is in the area of protecting printed circuit boards (PCBs), as shown in Fig. 8.5.

Electronic circuits are used to control all aspects of products from a marine engine ignition system to automobile air bag triggering mechanisms and from aerospace applications to digital watches. The PCB is usually comprised of a plastic circuit board substrate bristling with semiconductor, wires, and electric power supplies. The end-use application may require

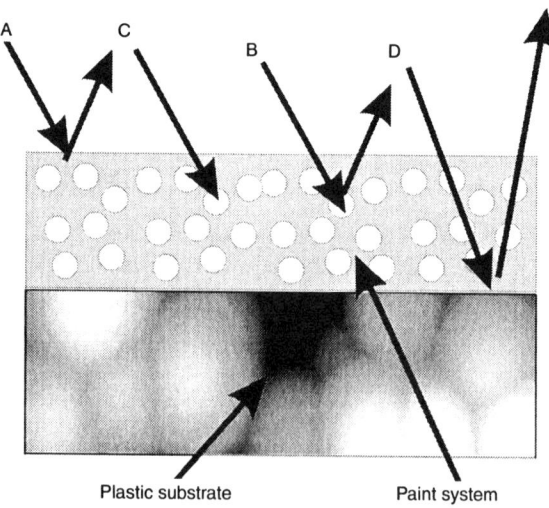

Fig. 8.4 Paint appearance and light

PCBs to experience temperature, dust, and humidity extremes. For example, a key electric circuit in a car that short circuits when driving through a puddle is certainly not desirable.

The PCBs can be conformal coated with a polymer that provides protection from the end-use environment and not require a special enclosure. These coatings are dielectric (insulating) polymers applied to the PCB by spraying, dipping, brushing, or vapor deposition polymerization. The coating creates an impenetrable boundary that matches (conforms to) the shape of the components being coated. Many of the conformal coatings are somewhat flexible and improve the impact and mechanical shock resistance of the PCB.

Conformal coating materials vary in chemistry, but these materials are the most popular: acrylic, polyurethane, epoxy, and silicone. Key attributes of the conformal coating materials include resistance to moisture and humidity, good electrical properties (resistance), excellent thermal properties, and resistance to attack from chemicals and microbials.

Application Techniques

There are several methods for applying paints to plastics. Spray painting is the method most frequently envisioned when considering painting; however, even spray painting has many variations that should be evaluated prior to selecting the best painting process for a particular application.

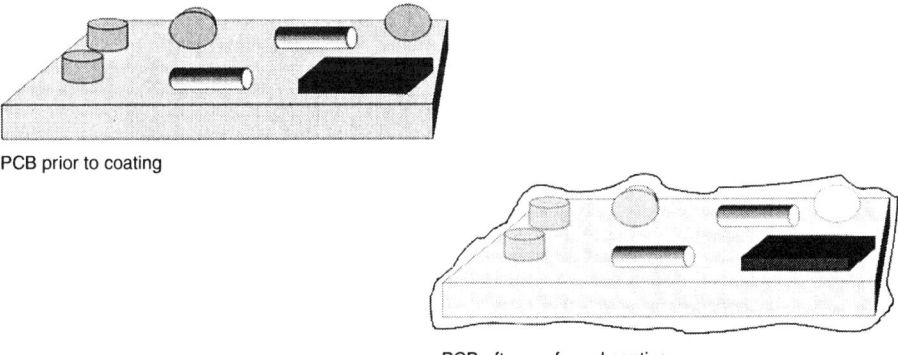

Fig. 8.5 Printed circuit board (PCB) conformal coating

Spray Painting Basics

Most people have some experience with spray painting, whether through hobbies/crafts using spray paint cans, home improvement using an airless sprayer to paint interior walls, or industrial painting of vehicles. All of these spray painting processes have basic concepts in common.

Spray painting, regardless of the specific method, reduces the paint to droplets, which are propelled toward the substrate being painted. When the droplets reach the surface of the substrate, they flow together to create a (hopefully) uniform paint film. As the binder (solvent or water) evaporates, the paint cures.

The size of the droplets is a function of the viscosity of the paint. The lower the viscosity is, the smaller the droplet size will be. Droplets are usually below 0.025 mm (0.001 in.) in diameter. Propelling the droplets should not introduce any foreign matter, including air, into the paint. The ability of the paint to flow on the surface of the substrate is a function of the viscosity of the paint, the surface condition of the substrate (see Chapter 9), and the chemical/physical interactions that may exist between the paint and the substrate.

Air spraying techniques are similar to using a can of spray paint, only air is the propellant. The air mixes with the paint at the nozzle. There is a significant amount of turbulence when the air and paint combine, which results in the paint being atomized (form droplets). The air stream carries the droplets toward the substrate (Fig. 8.6).

Airless. As the name implies, there is no air involved in airless painting. The paint in its storage vessel is pressurized under significant pressure (7 to 28 MPa, or 1 to 4 ksi). As the paint stream exits the nozzle, there is a significant reduction in pressure, which creates the droplets. The pressure reduction does not create as high a turbulent action at the nozzle compared to the air-based spraying, so the droplets tend to be larger.

Electrostatic

Electrostatic spray systems are innovative variants of conventional spray systems. Paint is atomized by an air or airless spray system. Additionally, an electrical charge is applied to the paint droplets as they form at the spray gun tip (Fig. 8.7). The charged particles tend to be uniformly separated as they travel toward the substrate. If the substrate is given an opposite charge to the paint particles, the paint droplets will be attracted to the substrate. This attraction significantly reduces overspray and waste. In some situations, a charged substrate may allow paint to wrap around the substrate, covering in a true three-dimensional (3-D) manner versus typical line-of-sight painting. An example of this 3-D painting is the coating/painting of chain-link fence. The paint is sprayed using an electrostatic gun from one side of the fence links, and the links are oppositely charged, which results in the paint wrapping around the metal links and painting all surfaces at the same time.

A closer look at the electrostatic process shows that there is an electric field between the spray gun and the substrate being painted. The voltage at the gun, about 90 kV, requires that the spray gun be relatively close (<30 cm, or <12 in.) from the substrate.

One key point, because plastics are excellent insulators and, therefore, difficult to charge, electrostatic spraying of plastic substrates has been limited. One area where plastics can utilize the advantage of electrostatic painting is where the plastic substrate has been plated or metallized prior to

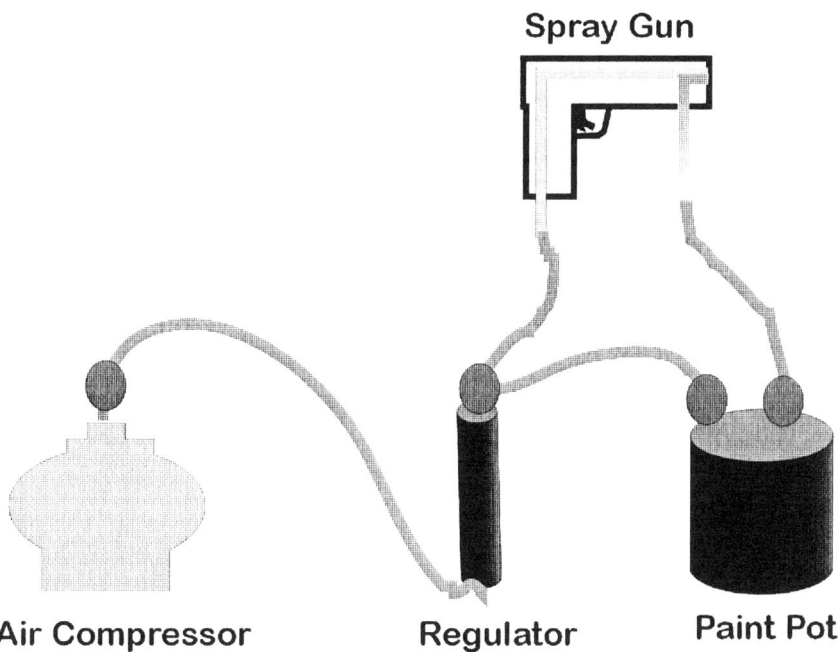

Fig. 8.6 Standard air-based paint spray system

192 / Decoration and Assembly of Plastics

painting. Automotive grills and wheel covers may be a candidate for this process.

Dipping

Dipping, as a paint application process, is exactly what the name implies. The substrate to be painted is literally immersed in a paint bath. The substrate is removed; the excess paint is drained; and the part is allowed to dry. The greatest advantage of dip coating is cost. There is little waste, and the equipment is relatively low in cost. The key to successful dip coating is proper substrate cleaning and preparation, as well as control of the paint viscosity.

Flow coating may be considered the opposite of dip coating. In the flow-coating process, the paint/coating material is poured over the substrate. The drained material is collected and reused. As with dip coating, there is little waste, and the equipment is relatively low in cost. The key to successful

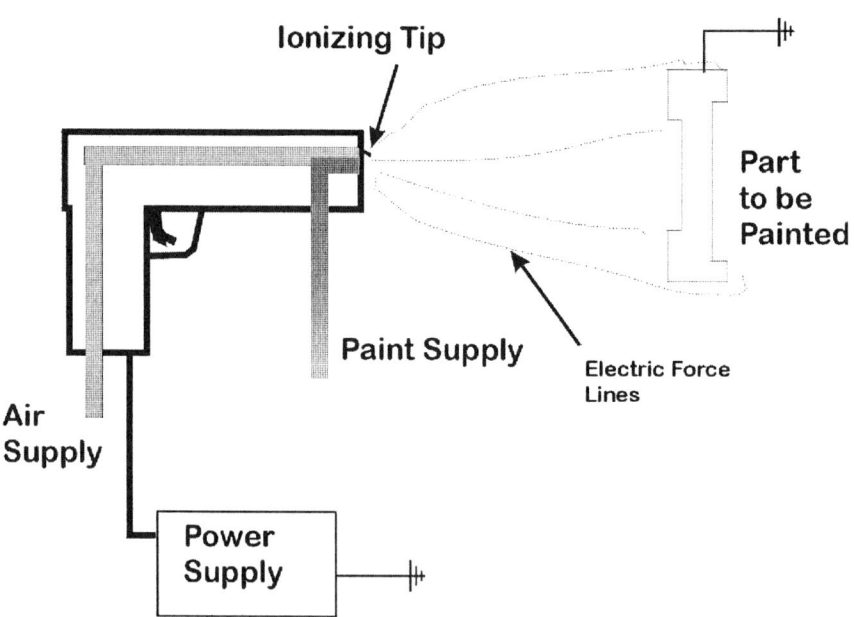

Fig. 8.7 Electrostatic painting

flow coating is proper substrate cleaning and preparation, as well as control of the paint viscosity.

Printing

Laser Printing/Etching

Laser usage in the plastics industry, especially in the area of decorating and assembly, has expanded significantly over the past 20 years. Laser printing of computer generated reports and graphics is now commonplace.

The use of lasers in plastics printing and etching is of particular interest because of its high speed and accuracy. Laser printing and etching can be accomplished within the production line and at production rates. There is no need to redirect production to a unique area to accomplish the laser-based symbolization.

Laser is an acronym for "light amplification by stimulated emission of radiation." Focusing its energy intensifies light energy. It is common knowledge that touching a common 100 W light bulb while on results in a burn. A 100 W laser, typical in industrial applications, is similar to taking the surface area of a 100 W light bulb and reducing the area so the light energy is focused through a 0.152 mm (0.006 in.) diameter. This energy results in an increase in heat several thousand times, hot enough to burn through plastic and metal.

Once this high energy is developed, the next step is to control it (Fig. 8.8). A generic set up integrates a computer to a laser. The computer is used to read a design file with the desired text or graphics. The computer translates the desired artwork to a series of servomotors and a moving (x-y) bed. The laser light is directed by the servomotor, which in turn controls a system of mirrors. This mechanism creates the effect of writing or drawing. The light passes through a lens to focus the beam onto the part. The plastic part to be printed or etched lies on a movable table that assists the laser writing. The result can either be a burn or etch. The burn creates an oxidized surface that can be an inkless printing. The etch, depending upon the depth, can also appear to be artwork or text. If the plastic substrate has multilayers, the etching can expose a lower layer of a different color that creates a more dramatic appearance.

194 / Decoration and Assembly of Plastics

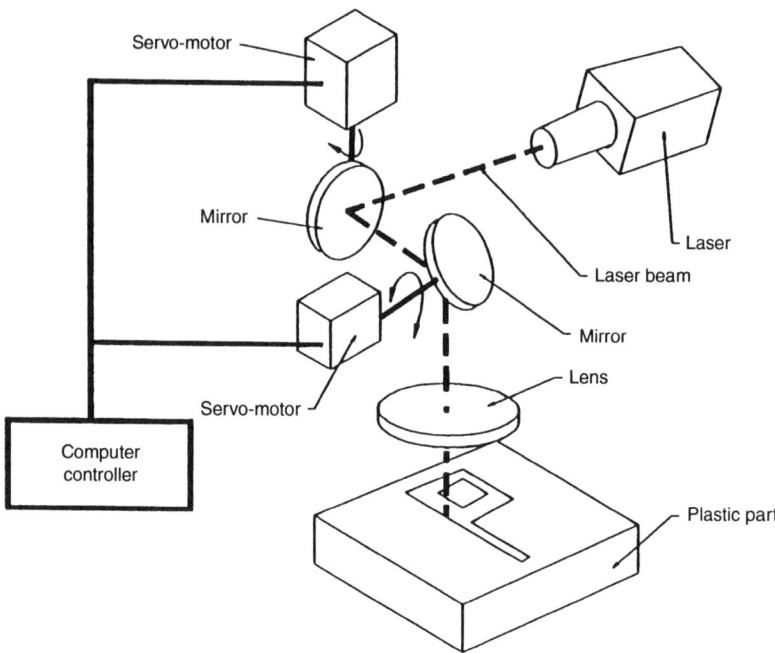

Fig. 8.8 Laser etching/coding

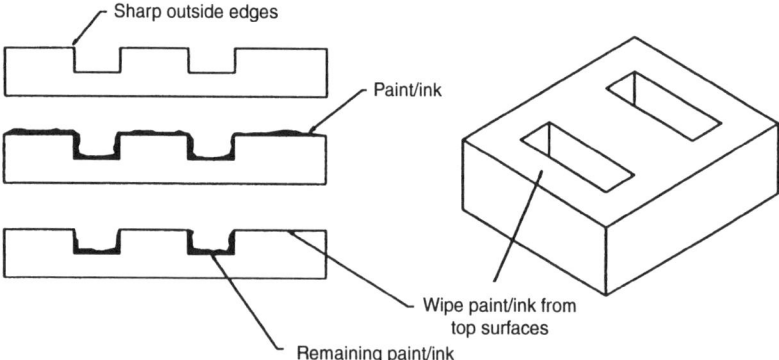

Fig. 8.9 Spray and wipe painting

Spray and Wipe

Spray and wipe is a process that requires the plastic part to have the desired code or image "cored" into the part so that the image lies below the actual surface of the part. Ink or paint is applied to the part, and the surface is immediately wiped off. The ink or paint in the cored area remains to highlight the code or artwork (Fig. 8.9).

TAFA

One area under development to complement the rapid prototype parts is TAFA (TAFA, Inc., Concord, NH). TAFA is the trade name for a metal

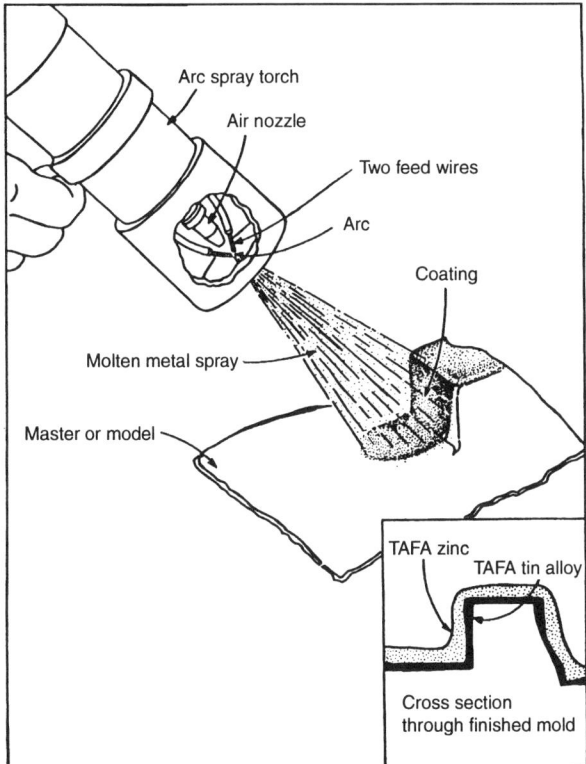

Fig. 8.10 TAFA arc spray moldmaking technique. Courtesy of TAFA, Inc., Concord NH

spraying process that has been available to the plastics industry for years, primarily as a metal spraying process for tool making. However, it is now finding a niche in the area of prototyping.

The TAFA process is capable of spraying an atomized metal coating quickly and uniformly over almost any geometry. The spray is developed in the TAFA gun (Fig. 8.10) and is extremely fine and atomized, thus allowing the metal coating to pick up the detail of the surface being coated. The gun utilizes a compressed air and electric arc metal feed as opposed to a flame spray. Because it does not use a flame spray, the TAFA process is kept below 66 °C (150 °F) and can be applied in a thickness over 1.587 mm ($\frac{1}{16}$ in.).

Screen Printing

Screen printing is a process that uses a finely meshed screen covered or filled with a photosensitive chemical (Fig. 8.11). The image to be reproduced is placed on the screen and exposed to light. The light cures the chemical not covered by the desired image, thus blocking the small holes in the mesh. The unexposed area is washed to expose the holes in the mesh. After the screen is made, the part to be printed is placed under the screen, and ink is forced through the mesh onto the part.

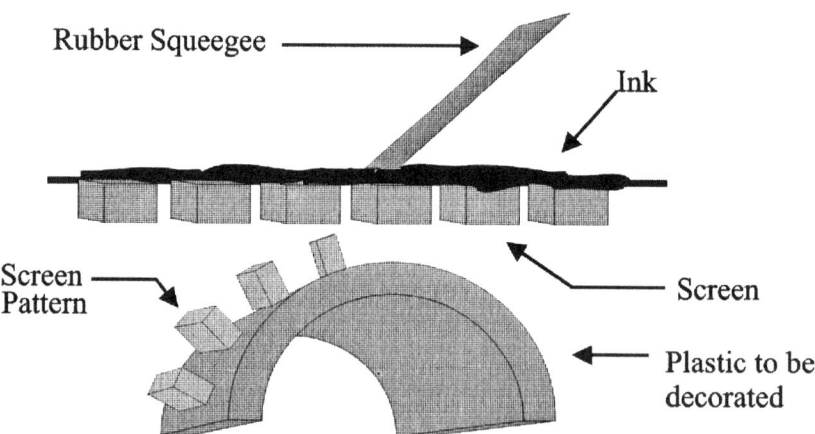

Fig. 8.11 Screen printing

Ink Jet Printing

Ink jet printing is a direct adaptation of the ink jet printer used to print documents. The ink jet concept actually sprays a crisp and highly directed focused beam of ink onto the plastic part. The printer has no direct contact with the part. The inks are fast setting and allow the designer to print date codes, part numbers, and messages on most plastics, regardless of the shape of the surface.

9

Surface Preparation

Plastic products are produced with a wide variety of plastic materials. Each plastic material has unique characteristics relative to surface properties. These property variations can affect adhesive bonding and adherence of decorative applications, as well as paint adhesion.

Surface preparation may be required to create the proper surface environment for durable decorating and assembly operations. Some of the key surface preparation techniques include corona discharge (widely used with polyolefin plastics, such as polyethylene and polypropylene), flame, plasma, and chemical.

Corona Discharge

The adhesive bond formation between paint, ink, or adhesive, and a substrate requires the establishment of interfacial molecular contact by wetting. Wettability is easily demonstrated by placing a liquid drop on a substrate and measuring the contact angle. Figure 9.1 shows typical liquid drop shapes to illustrate the difference between good and poor wettability. A liquid droplet set on a smooth, solid horizontal surface may spread out over the substrate, and the contact angle will approach 0° if complete wetting takes place. If wetting is partial, the resulting contact angle reaches equilibrium in the range of 0 to 180°.

Wettability is a function of one specific property of a solid surface—surface energy, often referred to as surface tension. Surface energy consists of two components: dispersion and polarity. Surface energy, like surface tension, is measured in dyne/cm. The higher the surface energy of the solid substrate relative to the surface tension of the liquid is, the better the wettability and the smaller the contact angle will be.

It is a rule of thumb that a good bond will form if the surface energy of a substrate exceeds the surface tension of the liquid by about 10 dyne/cm. Figure 9.2 shows absolute values of the surface energy of solid materials and the surface tension of liquids. Several polymers, such as polyethylene, polypropylene, and polystyrene, have inherently low surface energy and poor wettability, and adhesive bonding and finishing of parts made of these materials present problems to manufacturers.

The increasing use of waterborne primers, paints, and adhesives have made bonding more difficult. Water is a very polar liquid, and the substrate must be polar for water to wet the surface. In order for a waterborne chemical to uniformly coat the surface, the polar component of surface energy for the substrate should be at least 15 to 20 dyne/cm. To understand the magnitude of the problem, for example, the polarity of polyethylene and polypropylene are near zero. Surface treatment can increase surface energy and improve wettabililty and adhesive properties of polymer materials. The most versatile and effective surface treatment methods use electrical discharges in air.

Basics of Surface Modification with Electrical Discharges

In the presence of a high voltage field in the air gap, free electrons accelerate and ionize the gas, producing a gas plasma, which is an extremely reactive gas. The gas consists of free electrons, positive ions, and other

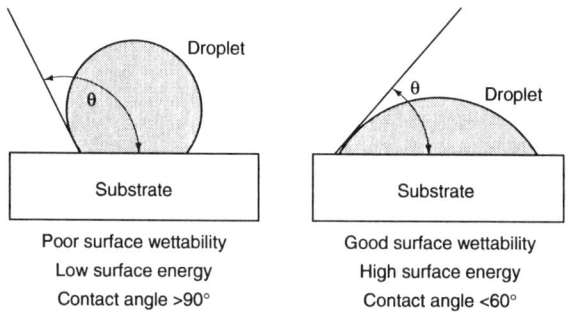

Fig. 9.1 Typical liquid drop shapes showing poor and good wettability

species. Plasmas exist over a wide range of temperatures and pressures. The solar corona, a lightning bolt, a flame, a spark in a spark plug, and a "neon" sign are all examples of plasma. The glow and arc discharges are notable examples of plasma at atmospheric pressure.

Free electrons, ions, metastables, radicals, and ultraviolet (UV) rays generated in the discharge impact the surface with energies sufficient to break the molecular bonds on the surface of most substrates (Fig. 9.3). This impact creates very reactive free radicals on the polymer surface. The free radicals can form crosslinks or, in the presence of oxygen, can react rapidly to form various chemical functional groups on the substrate surface.

Both auto adhesive and adhesive bondabilities can be greatly increased with the formation of polar functional groups in the electrical discharges. Functional groups produced on a polymer surface by the electrical discharge treatment include carbonyl (C = O), carboxyl (HOOC-), hydroperoxide (HOO-), and hydroxyl (HO-) groups. Even small amounts of appropriate reactive functional groups incorporated into polymers can greatly increase

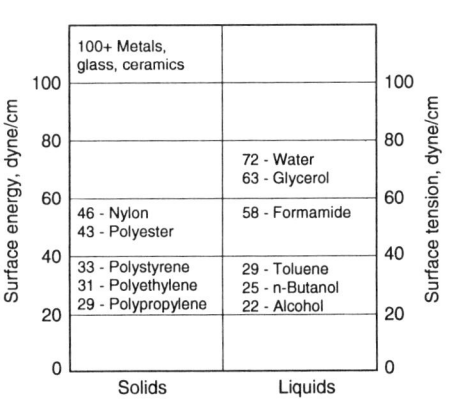

Fig. 9.2 Surface energy of solid materials versus surface tension of liquids

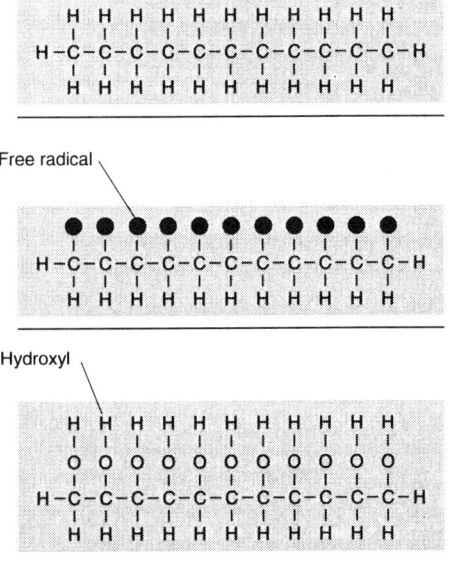

Fig. 9.3 Phases of surface modification

the adhesive bond strength with adhesives having high concentration of the functional groups. For example, polyurethane, epoxy, and acrylic adhesives contain hydroxyl. These adhesives are very effective for bonding to modified polyolefins, which after the treatment contain a large concentration of hydroxyl on the surface.

Electrical Discharge Treatment Equipment

Surface treatments with electrical discharges modify only the surface characteristics without affecting material bulk properties (Fig. 9.4). Several electrical discharges are used for surface treatment: arc discharge, spark breakdown, and glow discharge. The type of discharge depends on several controllable factors, such as electrode geometry, gas pressure, and power supply impedance. Key manufacturers have produced two types of equipment for treatment of three-dimensional objects: air-blown arc plasma treaters (spot treaters) and high-frequency arc treaters (electric surface treatment, EST, systems).

Air-blown arc plasma treaters are designed to separate gaseous plasma from a low-frequency electrical arc. A treating head in these devices contains two diverging wire electrodes and a fan blower. An electric arc is established between the electrodes. The fan blows air through the arc and

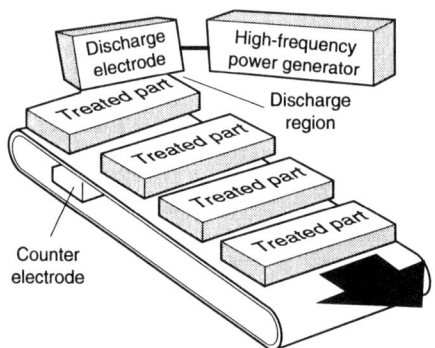

Fig. 9.4 Surface treatment with electrical discharges modifies only the surface characteristics without affecting material bulk properties

produces a plume of plasma gas, which extends past the electrodes to the surface to be treated.

High-frequency arc treaters (EST systems) use high-frequency arc discharge for surface treatment. The discharge is generated between two electrodes located on opposite sides of the treated surface. Parts are treated as they move through the treating area (Fig. 9.4). The intense discharge is sustained in a large air gap by establishing a high potential difference between the electrodes. This arrangement is very effective, because practically all the energy produced in the discharge is applied to the treated surface. Special insulation on one or both electrodes assures stable, uniform, and continuous discharge. For larger parts, one of the electrodes is specially shaped so the part fits over it.

High operating voltage is only one condition for effective treatment. Higher efficiency of energy transfer is achieved at high frequencies. It has been shown that discharges at higher frequencies lower the power required to achieve a given treatment level.

The EST systems utilize high impedance ac power generators and high voltage transformers, which produce up to 70,000 V at frequencies from 20 to 30 kHz. The high voltage and frequency set this technology apart from conventional corona technology, which operates at lower frequencies and/or voltages. The combination of high frequency and voltage makes some advanced equipment suitable for treatment of three-dimensional parts at high line speed.

Applications for EST include the following:

- Polystyrene labware for culture grade products
- Surfaces of biomedical testing devices for improved wettability
- Automotive exterior and interior trim parts for painting, adhesive bonding, and foam application
- Plastic housings of electronic components for improved adhesion before encapsulation and product marking
- Plastic bodies of electronic connectors for increased pull force of connector pins
- Syringe barrels before printing
- Needle hubs prior to bonding stainless steel needles
- Electronic cable insulation for improved adhesion of inks and coatings
- Lids and covers of chemical containers for gasket application
- Automotive profiles made of ethylene-propylene-diene monomer (EPDM) rubber for flocking and coating

Other Treatment Issues

Surface Cleanliness. The quality of treatment depends to a great degree on surface cleanliness. The surface must be absolutely clean for proper treatment to occur. The following are some surface contaminants that reduce treatment efficiency: additives, mold releases, machine oil, and grease.

Shelf life of treated surfaces ranges from hours to months, depending on the plastic, the formulation, and the ambient environment of the storage area. Polymer chain mobility in the treated material causes the bonding sites to move away from the surface. The higher the chain mobility is, the faster the aging of the treatment will be. Exposure of treated surfaces to elevated temperature increases the chain mobility. Material purity is an important factor. Shelf life of treated parts is limited by the presence of low molecular weight components, such as antiblock agents and amides. Eventually, these components migrate to the polymer surface. It is, therefore, recommended to apply an adhesive or coating soon after the treatment.

Figure 9.5 shows typical aging characteristics of a treated polyolefin. Surface energy decreases after the treatment and then stabilizes at a level higher than the initial surface energy. Treatment on surfaces of most polyolefins last from hours to days. Figures 9.6 to 9.8 show the electrical surface treatment of polypropylene inner door panels.

Flame Treatment

While gas flame treatment offers many advantages and has been in existence for several decades, until recently, it has invariably taken a back seat to

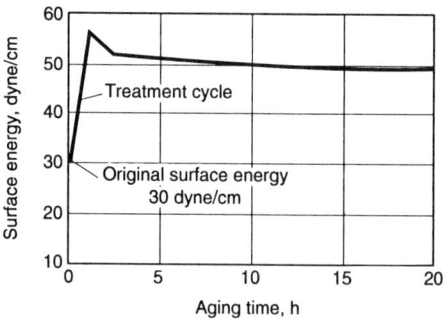

Fig. 9.5 Typical aging characteristics of a treated polyolefin

Fig. 9.6 Three-dimensional plastic (polypropylene) interior door handles. Courtesy of Tantec, Inc.

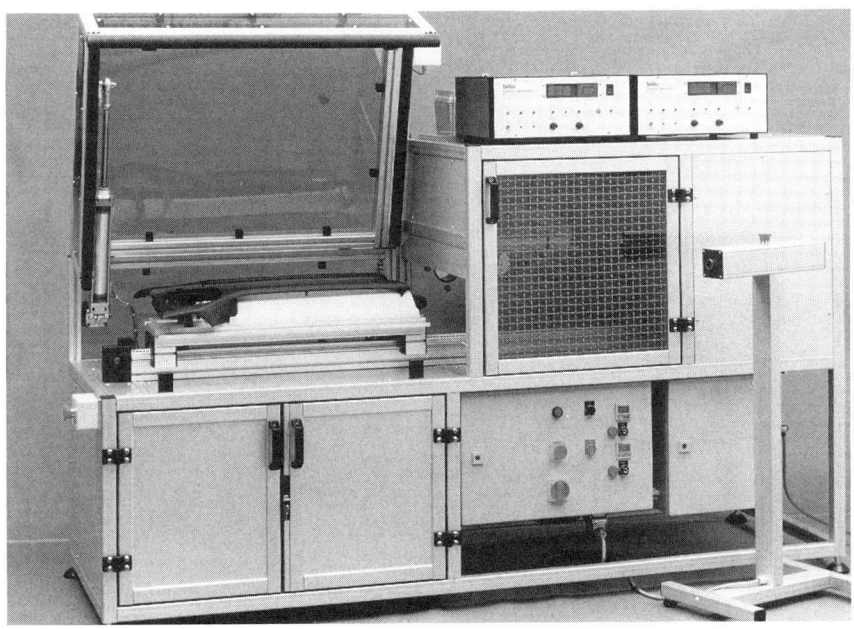

Fig. 9.7 Plastic parts being loaded into electric surface treatment apparatus. Courtesy of Tantec, Inc.

other methods of surface treatment. However, the move toward high-performance polymer blends that call for better quality printed, coated, and laminated products and the environmental pressure to change to water-based inks and adhesives have sparked a resurgence in the popularity of flame treatment and the benefits that can be derived. A better understanding of the process and new technology have given rise to advanced control systems and treatment stations that provide consistent and superior surface treatment.

The benefits of controlled gas flame treatment of plastics include the following:

- Higher treatment levels
- Low treatment decay rates
- Guaranteed no backside treatment
- Suitable for both webs and profiled objects
- No off taste
- No ozone
- No film odor
- Surface decontamination of film, foil, and paper

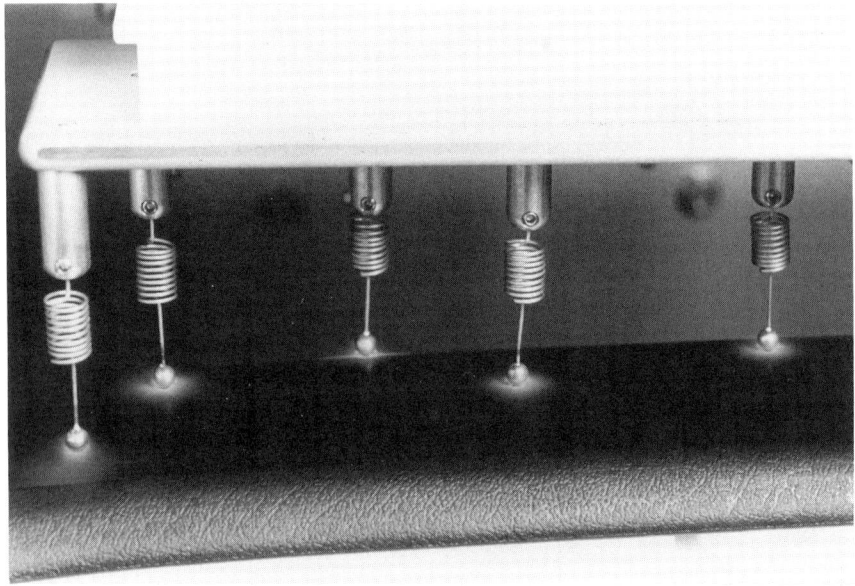

Fig. 9.8 Close-up of electrical surface treatment process. Courtesy of Tantec, Inc.

- Fiber removal
- No material gauge restrictions
- No pin holing
- No reel blocking

The naturally low surface energy of most polymer products and the inert nature of other substrates, such as rubber, foil, and paper, invariably demand surface treatment to promote a bond with ink, paint, coatings, adhesives, and laminates. This need for treatment arises where the surface energy of the substrate is less than the surface tension of the product being applied.

Figure 9.9 shows how solids and liquids differ in their natural surface properties. A liquid applied to a substrate with a relatively lower dyne level is inclined to retain a droplet form (Fig. 9.10). Once treated, the increased surface energy of the substrate offers an attractive molecular structure, which overrides the liquid surface tension, promoting "wet-out" onto the substrate and an improved bonding between the two materials (Fig. 9.11).

Gas flame treatment not only provides a highly receptive surface for printing, coating, and laminating, but also the exothermic reaction created in the gas combustion produces an ambient flame temperature of 1800 °C, which decontaminates impurities and removes surface fibers in paper and fabric. Individual processes have differing treatment level requirements:

Purpose	Treatment level, dyne/cm
Flexo printing	<40
Gravure printing	<43
Litho/offset printing	<43
Water-based ink printing	46–50
Screen printing	<43
Laminating	Normally, no upper limit
Solvent-based	44–48
Water-based	54–58
Gluing	>45
Coating	>45 (usually up to 54)

The following assessment of flame treating for printing is based on heavy solvent-laden inks. As industry moves toward water-based liquids, the need for high levels of surface energy obtainable from flame treatment becomes obvious.

208 / Decoration and Assembly of Plastics

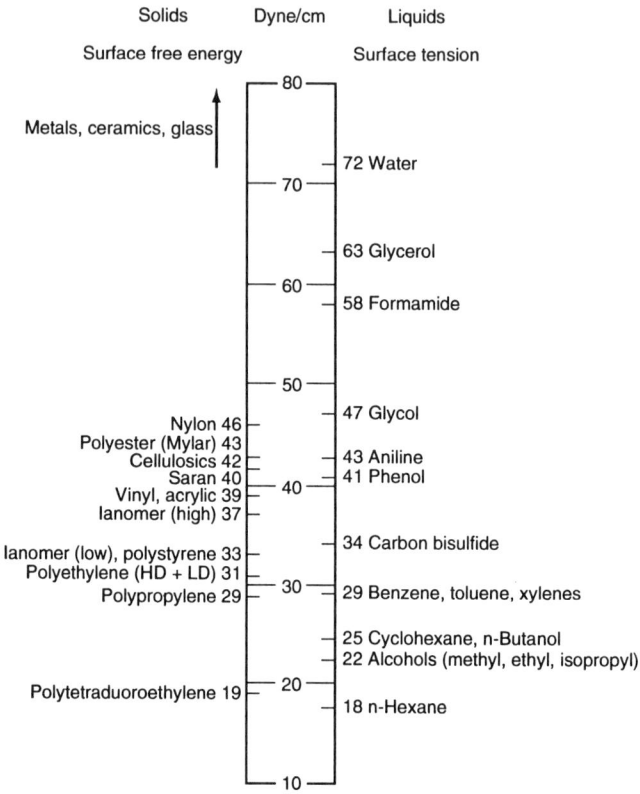

Fig. 9.9 Comparison between solid surface free energies and liquid surface tension

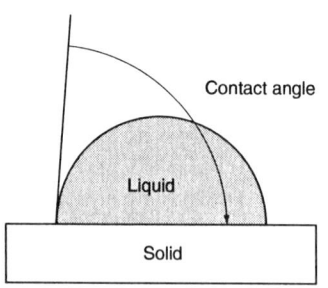

Fig. 9.10 Contact angle of liquid droplet on untreated solid surface

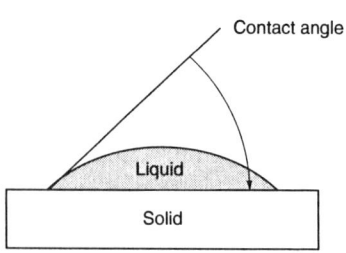

Fig. 9.11 Contact angle of liquid droplet on treated solid surface

Theory

Studies show the increase in surface energy and bond strength created by flame treatment is derived through the application of excited chemical species and electron energies. A gas flame is the combustion between a fuel (methane, propane, or butane) and an oxidizing element (air). This combustion produces a complex exothermic reaction, or plasma, during which oxygen molecules disassociate into free oxygen atoms that bombard the material surface (Fig. 9.12).

The absolute plasma includes many energized species: free radicals, ions, neutrals, and electrons (O, OH, NH, NO, and CN). However, conclusive evidence suggests it is the oxygen content that is most significant (Fig. 9.13).

In addition to the chemical properties, the plasma shows electron energies of 0.5 eV. While this may be considered low, the high mass flow rate of a flame system will polarize the polymer to depths greater than a single molecular layer. For example, for an electron density of 10^8/cm and an electron energy of 0.5 eV, the polarization penetration will be less than or equal to several nanometers. The combination of these chemical and polar charges provide a highly receptive surface on a host of substrates for all coating, painting/printing, and laminating operations.

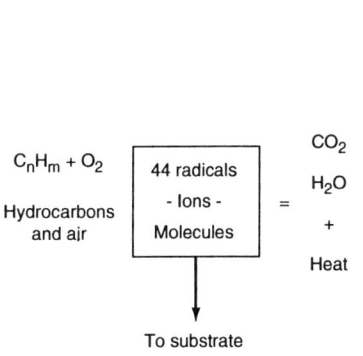

Fig. 9.12 Complex exothermic reaction caused by combustion between a fuel and an oxidizing element

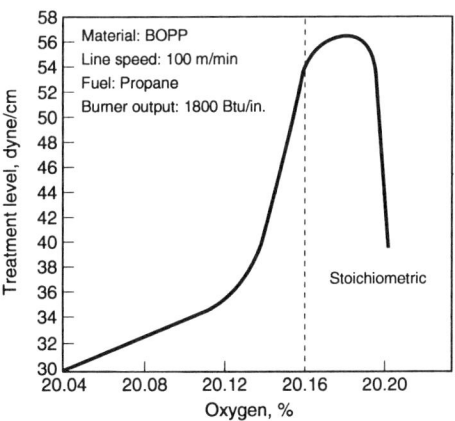

Fig. 9.13 Treatment level versus oxygen in premix

The key to optimal treatment levels requires incorporating an understanding of the process variables into the control factors. The chemical and electron energy properties are crucial to the process and require precise control. The intensity and volume applied to the substrate depend on the following conditions: air to gas ratio, flame geometry, and flame-to-material interface.

Gas/Air Mixture Control (Fig. 9.14)

Optimal flame conditions are achieved when combustion reaches the stoichiometric condition, whereby sufficient oxygen is supplied to combust all of the fuel gas chemicals in the supply mixture.

The conventional means of producing and maintaining air/gas mixture ratios utilized a venturi mixer combined with a zero gas regulator. Combustion air at the required pressure passes into the mixer where it develops maximum velocity. The kinetic energy created by this velocity air stream in the venturi throat creates a negative pressure at the entrance of the gas port. This negative pressure draws gas from the zero pressure regulator to produce the required air/gas mixture ratio.

Gas	Air to gas ratio
Methane (CH_4)	10:1
Propane (C_3H_6)	25:1
Butane (C_4H_{10})	32:1

This system works satisfactorily at the commissioning stage. It also works for systems that do not require precise control of the air/gas ratio. However, two problems exist. First, changes in ambient temperature to the venturi mixer will alter an air/gas ratio from an optimal setting. A 15 °C rise is

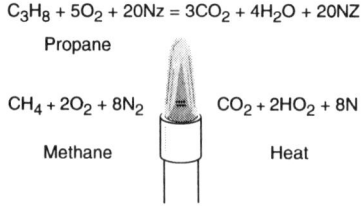

$C_3H_8 + 5O_2 + 20N_2 = 3CO_2 + 4H_2O + 20N_2$
Propane

$CH_4 + 2O_2 + 8N_2$ $CO_2 + 2HO_2 + 8N$
Methane   Heat

Fig. 9.14 Air/gas mixture control, optimal gas combination

sufficient to produce a 3% change in the air/gas ratio resulting in a 14 dyne/cm drop in surface treatment. Second, changes in gas composition that occur in both commercial and main gas give rise to an imbalance in the set ratios and oxygen content. These inherent problems are overcome by the use of an oxygen analyzer, which samples the post mix gases. The oxygen content can be directly related to the flame temperature (Fig. 9.15), ionizing current (Fig. 9.16), and ultimately, treatment level (Fig. 9.13). Advanced manufacturers of gas treatment systems have decided to sample the postmix gases as opposed to the premix gases. This practice takes into account the variable content of gas supplies as well as atmospheric oxygen concentration (Fig. 9.13).

The analytic system simulates what happens at the material face by burning the fuel gases and measuring the residual oxygen by a zirconia cell. Then, a feedback signal proceeds to a motorized control valve, which adjusts the gas input accordingly (Fig. 9.17). Using the analyzer in conjunction with the venturi system maintains the ratio required to achieve the stoichiometric optimal flame conditions.

Flame Geometry

Treatment levels do not depend on only the correct gas but also on how the mixture is released for combustion. Different processes require different thermal ratings: film and polymer products, 5,000 Btu/h/in.; metal, foil, and

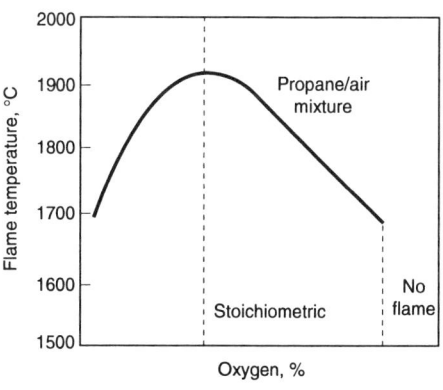

Fig. 9.15 Flame temperature versus oxygen in premix

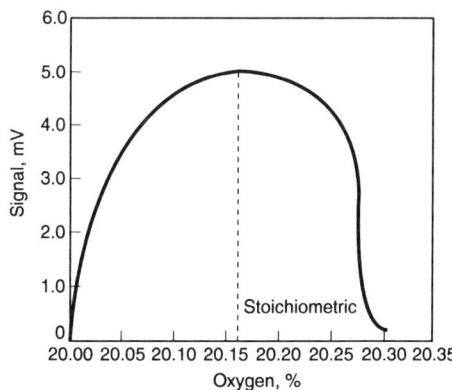

Fig. 9.16 Flame ionization current versus oxygen in premix

film perforation, 10,000 Btu/h/in.; and paper and board coating, 25,000 Btu/h/in. Key manufacturers have designed burners to provide a stable regulatory platform that supports flame pattern and offers even treatment across the width of the substrate.

An examination of the flame structure (Fig. 9.18) reveals that it has two zones. The primary (or reducing) zone is an area of incomplete combustion and is the coolest region of the flame. The secondary (or oxidized) zone is the region that defines the area of complete combustion containing the excited species and is the hottest region of the flame.

The positioning of the substrate in the flame is an important variable. The optimal distance depends on other factors, such as line speed and the Btu output of the burner (Fig. 9.19). The actual treating portion of the flame also depends on the Btu output of the burner, which depends on burner port size. The surface should not come in contact with the inner cone of the flame, because this is the reducing zone and will weaken treatment. Generally

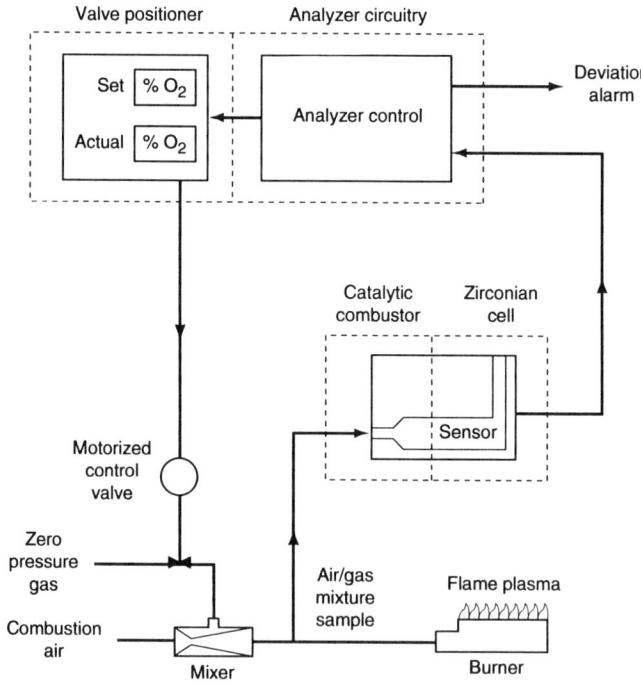

Fig. 9.17 Oxygen analyzer and feedback system

adequate treatment of most materials will be achieved with spacing of 5 to 15 mm between the material and the tips of the cones.

The shape and relative sizes of the two cones depend on the composition of the two gases (the air/gas ratio). Studies have shown that optimal treatment on the surface of a polyolefin is just above the stoichiometric value (i.e., air in slight excess to that required for complete combustion), as shown in Fig. 9.20.

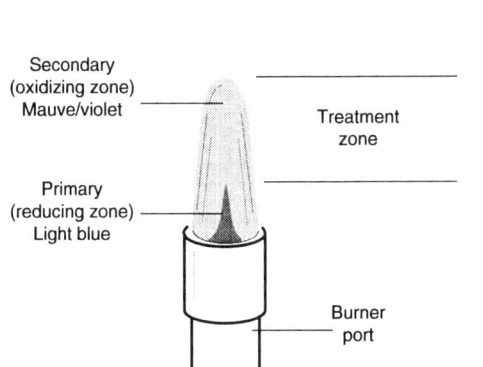

Fig. 9.18 Flame structure

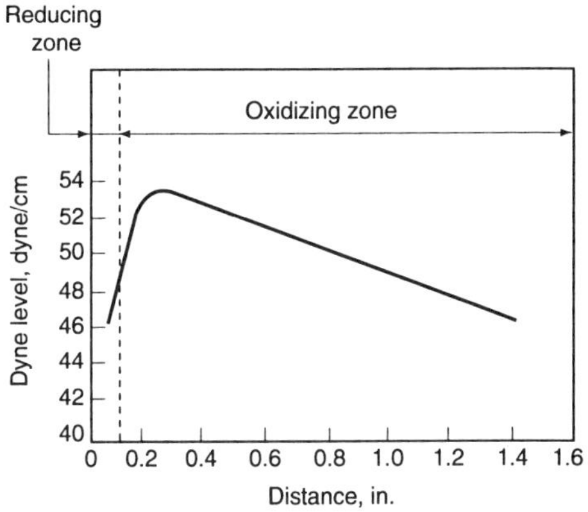

Fig. 9.19 Btu output of a burner

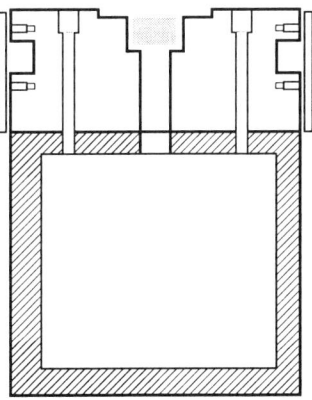

Fig. 9.20 Section through water-cooled burner

For ease of maintenance and for replacement, the ribbon pack can be removed.

Electronic Control

Because of the unique demands and variability of installation, the requirements of an out system of burner ignition and flame supervision are largely manual, while still retaining all the normal interlocks and logic sequences associated with more standard automatic systems.

Combustion, extraction, and pressurization fans (where fitted) are individually push-button started, and satisfactory operation is monitored and interlocked using pressure switches. The ignition of the burner is a push-button operation that initiates a high-temperature ignition spark at one end of the burner while opening the gas safety shut-off valves. The flame is detected by a rectification system via a second electrode at the far end of the burner. When the flame is detected, the ignition spark is stopped immediately.

The burner ignition method is preferable to automatic systems for the following reasons. Chamber purge is usually not applicable, because the burner is not enclosed in a chamber. In some instances, the distance between the mixing system and the burner is excessive and renders normal ignition/detection times inadequate, but push-button ignition overcomes this. Because burners are not normally enclosed, there is no overwhelming need to limit gas release during ignition (there can be no build up). Where the burner is substantially enclosed (e.g., in a flame proof/pressurized enclosure), the same system is used. However, timers are incorporated into the circuit that ensure adequate prepurge, limit the trial for ignition period, and also prevent a reignition attempt until a repurge time has elapsed. Thus, safe operation and the flexibility demanded of the ignition system are maintained.

Once running the burner, throughput is subject to whatever manual or automatic systems are applied, but it is always monitored by the flame supervision unit. If the flame ceases to be detected or any of the pressure switches or other interlocks (e.g., water flow) drops out, the burner shuts down immediately.

The question of safety often arises when discussing gas flame equipment. Advanced gas treatment units are manufactured to rigorous British, European, and international safety standards. Systems are fully interlocked with guards where necessary. More recently, some advanced manufacturers have supplied a pressurized unit for installation onto a solvent coating line, overcoming the obvious solvent hazards.

Plasma

Historically, treatments to improve the adhesion of coatings to plastics consist of mechanical abrasion, solvent wipe, solvent swell followed by acid or caustic etching, and corona. Each of these treatments has limitations, providing a strong driving force for alternative treatments. Often what works for one plastic will not be effective for another; thus, specific treatments need to be developed for each.

Mechanical abrasion or sand blasting is operator sensitive, operator intensive, dirty, and difficult on small parts or high-production volumes. Grit blasting also presents both occupational health and safety and environmental risks. Solvents pollute, present safety and expensive disposal problems, and often do not work. Acid etching, although more effective than solvent swelling, usually compounds handling problems, and it is easy to overtreat and damage parts.

Corona treatment is limited both to materials that are responsive to treatment and part configuration. Often the treatment is marginal in effectiveness and short-lived. Because a corona relies on ambient air, the process can change from day to day and from location to location. Complex shapes cannot be easily treated, because the treatment quality is a function of the distance to the electrodes. Arcing is prevalent in atmospheric discharge systems (corona) and can damage the treated material either thermally or electrically. Effluent from a corona line must be treated to eliminate the ozone the process generates.

Plasma Surface Treatment

Plasma is a partially ionized gas containing ions, electrons, and various neutral species at many levels of excitation. The free electrons gain energy from an imposed electric field, colliding with neutral gas molecules and transferring energy. The collisions and transfer of energy form free radicals, atoms, and ions, which interact with solid surfaces placed in the plasma. This interaction leads to drastic modifications of the molecular structure, providing the desired surface properties (Fig. 9.21).

The effect of a plasma on a polymer is determined by the chemistry of the reaction. The plasma process causes changes only to the surface of the material, several molecular layers deep. In addition, plasma changes the molecular weight of the surface layer by scissioning (reducing the length of molecules), branching, and crosslinking. The type of surface change that occurs depends on the composition of the surface and the gas used.

One of the more common plasma processes to enhance adhesion is treatment in a cold gas oxygen plasma. Oxygen plasma is aggressive and forms numerous components. Species found within an oxygen plasma are O^+, O^-, O_2^+, O_2^-, O, O_3, ionized ozone, metastable-excited O_2, and free electrons. As the components recombine, they release energy and photons, emitting a faint blue glow and much UV radiation. The photons in the w region have enough energy to break the carbon-carbon and carbon-hydrogen bonds:

$$RH + O^* \rightarrow R^* + OH^*$$

$$RH + O^* \rightarrow R'' + R''O^*$$

$$RH + 2O \rightarrow R^* + H^* + O_2$$

$$RH \stackrel{UV}{\rightarrow} R^* + H^*$$

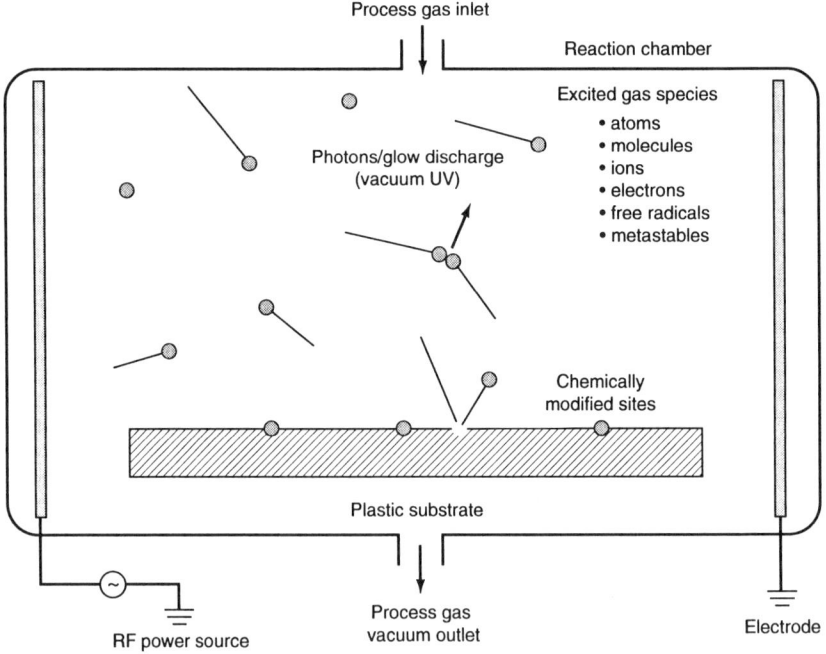

Fig. 9.21 Surface modification of plastic in a gas plasma reactor

All of the active species react with the polymer, in addition to bombardment by photons, ions, and neutral particles. The byproducts are CO_2, CO, H_2O, and hydrocarbons with low molecular weight that are readily removed by the vacuum system. These molecules can be excited by the radio frequency (RF) field, but the effect on the reaction appears insignificant when used in an oxygen plasma.

Tetrafluoromethane added to the process gas greatly increases the efficiency of an oxygen plasma. The fluorine is more effective at breaking the carbon-carbon and carbon-hydrogen bonds, which is the limiting reaction in an oxygen plasma. The O_2/CF_4 plasma yields excited forms of O, OF, CF_3, CO, CO_2, and F, which react with the hydrocarbon to produce HF and H_2O.

The selection of the process gas determines how the plasma will alter the polymer. Very aggressive plasmas can be created from relatively benign gases. For example, a tetrafluoromethane (Freon [E.I. Du Pont de Nemours & Co., Inc., Wilmington, DE]) plasma contains free radicals of fluorine. Oxidation by fluorine free radicals is known to be as effective as oxidation by the strongest mineral acid etchant solutions with one important difference: the plasma byproducts do not require special handling. As soon as the plasma is shut off or the excited species exit the RF field, the species recombine to the original stable and nonreactive form within a few seconds.

Gases or mixtures of gases used for plasma treatment of polymers include nitrogen, argon, oxygen, nitrous oxide, helium, tetrafluoromethane, water, and ammonia. Each gas produces a unique plasma chemistry. Surface energy can be increased very quickly by plasma-induced oxidation, nitration, hydrolysis, or amination. Gases that contain oxygen are generally more effective at increasing surface energy. For example, plasma-induced oxidation of polypropylene increases the initial surface energy of 29 mN/m to well over 73 mN/m in just a few seconds. At 73 mN/m, the polypropylene surface is completely water wettable.

Three competing molecular reactions alter the plastic simultaneously, and the extent of each depends on the chemistry and the process variables:

- *Ablation:* Removal by evaporation of surface material
- *Crosslinking:* Connecting of two or more adjacent polymer chains
- *Activation:* Substitution of atoms in the molecule with chemical groups from the plasma

Ablation is an evaporation process. The plasma breaks covalent bonds of the polymer backbone by bombardment with high-energy particles, such as free radicals, electrons, and ions. As long molecules become shorter, volatile oligomers and monomers boil off (ablate) and are swept away with the exhaust.

Crosslinking, or unsaturation, on the other hand, is carried out with an inert process gas (oxygen-free argon or helium). The bond breaking occurs on the polymer surface as with the oxygen plasma. Because there are no free-radical scavengers (oxygen byproducts) in the plasma, the molecule can do one of three things. It can recombine with the byproducts and revert back to the original state; react with an adjoining free radical on the same chain, forming a double or triple bond; or form a bond with a nearby free radical on a different chain (crosslink).

In the activation process, surface polymer groups are replaced with atoms or chemical groups from the plasma. The plasma breaks the backbone of the polymer, or the pendant atoms or groups from the backbone, creating free radicals on the surface. Being thermodynamically unstable, the free radicals reach out in the environment (the plasma) to couple with free-radical species forming stable, covalently bonded atoms or more complex groups. The new groups on the polymer surface alter the adhesion characteristics. Activation reactions are best completed after a cleaning plasma to ensure that the surface is free of contamination.

Depending on the chemistry of the polymer and the gas, substitution can make polymers wettable or totally nonwettable. The type of atoms or groups substituted determine how the surface will behave. If the plasma creates free-radical polar groups, such as hydroxyl from a water vapor plasma treatment, the surface becomes wettable. Substitution can also be used to create a nonwettable barrier layer that inhibits chemical penetration. Fluorine-containing plasmas produce a nonwettable surface by reducing the surface energy. The hydrocarbon chain acquires the properties of a fluorocarbon. In theory, either fluorine or trifluoromethyl radicals substitute for abstracted hydrogen. Electron spectroscopy for chemical analysis (ESCA) verifies this hypothesis.

Such treatment is important for polymers used in medicine, where it is often undesirable to have catheters wetted by blood. The major advantage is that such modification can be accomplished using innocuous gases, such as tetrafluoromethane, instead of reactive and hazardous fluorine. Plasma contains very high-energy vacuum UV radiation, which is often underappreciated in polymer processing. Wetting creates free radicals on the surface of the polymer. Such radicals are identical in nature to radicals created by electron bombardment, and they react in a similar manner to chemically modify the surface.

Effectiveness of Plasma Processes

Treating a polymer with plasma can increase the surface energy by modifying the surface chemistry. Greater surface energy translates to greater chemical reactivity and compatibility with adhesives, paints, inks, and deposited metallic films. The enhanced surface reactivity is characterized in the laboratory by water wettability. Wettability describes the ability of a liquid to spread over and penetrate a surface, and it is measured by the contact angle between the liquid and the surface. The relationship between contact angle and surface energy is direct: contact angle decreases with surface energy.

Contact angle measurements are sometimes used as a general indication of the presence of contaminants. The cleaner the surface is, the lower the contact angle of a water drop will make with the surface. For example, a surface contaminated with mold release will make a contact angle of 80 to 90°, indicating poor wettability, and silicones will form a contact angle greater than 90°. Many clean metal surfaces show a contact angle of 30 to 70°. On the other hand, most plasma-treated surfaces yield a contact angle of 20° or less and, therefore, assure improved adhesion.

Bonding in manufacturing processes is a specialized field, but generally cleanliness and wettability are necessary for good adhesion. Plasma cleaning and activation greatly increase the apparent bonded surface area. Scanning electron micrographs have verified the increase in surface area when protracted plasma etch times (1 min or longer) are employed.

High surface energy alone does not guarantee better adhesion. Plasma is a chemical process, and the results of the operation depend on the chemistry of the surface and the plasma. The resultant surface chemistry must be compatible with any bonding agents, inks, paints, or metals applied. Fortunately, gas plasma allows the surface to be reengineered for optimal performance, providing the manufacturer tremendous versatility in materials, design choices, and materials costs.

Fluoropolymers are a group of polymers in which part or all of the hydrogen has been replaced by fluorine. In general, fluoropolymers have outstanding temperature and chemical resistance. These properties make fluoropolymers difficult to modify for bonding or surface decoration. The only previous option was to use an etchant based on metallic sodium, which poses handling and disposal problems.

Treatment in plasma is an effective method of modifying fluoropolymer surfaces. Plasma is inexpensive, clean, safe, and has no waste products that need special disposal. The fluoropolymer surface altered by plasma is completely wettable, can be printed on, and can be bonded by structural adhesives.

Table 9.1 shows the improvement in peel strength after treatment in a sodium-based etchant and after treatment in an oxygen plasma.

Other Polymers. A more extensive study on a copolymer of ethylene and tetrafluoroethylene, Tefzel 200 $[-(CH_2-CH_2)_m \sim (CF_2-CF_2)_n-]$ (E.I. Du Pont de Nemours & Co., Inc., Wilmington, DE), compared sodium etch with oxygen-plasma treatment. The plasma-treated specimens exhibited two to three times greater lap shear adhesive strength than Tetra-Etch (W.L. Gore & Associates, Inc., Fluoropolymer Etchants and Services Group, Flagstaff, AZ) treated specimens. An ESCA was used to verify the addition of oxygen, as well as the type of oxygen groups created on the Tefzel surface as a result of the plasma treatment.

Engineering Plastics

Plasma processes provide a facile method to obtain quality adhesive bonds with engineering plastics, such as polycarbonates, polyacetals, aromatic esters, imides, amides, and liquid crystal polymers. After plasma treatment, the mode of failure often changes from adhesive (interfacial) to cohesive in the adherend.

In a previous study, the effects of oxygen and ammonia plasmas on engineering plastic specimens bonded with two adhesive systems were compared. The adhesives employed were Scotch-Weld 3549 (3M, St. Paul, MN) urethane and Scotch-Weld 2219 (3M, St. Paul, MN) epoxy, both two-component, room-temperature-curing adhesives. The engineering plastics were supplied by the manufacturers as injection molded, dog-bone style, tensile test specimens. Each specimen was cut midway through the gauge length, plasma treated (except the control), and bonded to provide a 12.7 by

Table 9.1 Peel strength of adhesive-bonded fluoropolymers

	Peel strength, N/m	
	PFA(a)	FEP(a)
Untreated	(b)	17.5
Chemically etched	1120	1430
Plasma treated	1450	1820

(a) PFA, a perfluoro alkoxy, and FEP, fluorinated ethylene-propylene, are members of the Teflon range of products from E.I. Du Pont de Nemours and Co., Inc. (Wilmington, DE). (b) Too low to measure

12.7 mm overlap. Lap shear tests were conducted on a universal testing machine in a tensile mode, using a cross-head speed of 12.7 mm/min (values are the average of at least three specimens at each condition). Control specimens (not plasma treated) were solvent wiped with ethanol to remove contaminants and air dried prior to bonding. Table 9.2 presents the results for the epoxy bonding study. The results demonstrate that one gas was not the best choice for all cases, but rather a complex interaction occurs in the chemistry of the base polymer, the plasma, and the adhesive.

Commodity Resins

Plasma processing, while a leading-edge technology, is not limited to engineering and sophisticated resins. As early as 1969, researchers reported that plasma treatment of polyethylene and Nylon 6 adherend surfaces greatly improves bondability, with bond failures often occurring in the polymer rather than at the adhesive/polymer interface (Table 9.3). The specimens

Table 9.2 Lap shear strength of untreated and plasma-treated surfaces

Material	Manufacturer	Type of material	Preferred plasma gas	Lap shear strength, MPa Control	Plasma	Improvement	Failure mode
Valox 310	General Electric	Polyester thermoplastic	O_2	3.6	11.3	3.1×	From adhesive to material
Noryl 731	General Electric	Polyphenylene ether	NH_4	4.3	12.4	2.9×	From adhesive to material
Durel	Hoechst Celanese	Poyarylate	NH_4	1.7	14.9	8.6×	From adhesive to cohesive
Vectra A625	Hoechst Celanese	Liquid crystal polymer	O_2	6.5	8.6	1.3×	From adhesive to material
Delrin 503	Du Pont	Acetal homopolymer	O_2	1.1	4.5	3.9×	From adhesive to cohesive
Ultem 1000	General Electric	Polyetherimide	NH_4	1.3	14.4	11.3×	From adhesive to cohesive
Lexan 121	General Electric	Polycarbonate	O_2	11.8	15.5	1.3×	From adhesive to cohesive

Table 9.3 Improved bond strength following plasma treatment

Polymer	Treatment	Bond strength, MPa Average	Low	High
HDPE	None	2.2	1.8	2.5
	He plasma	21.6	21.0	22.3
	O_2	17.4	17.0	18.0
Nylon 6	None	5.5	4.4	7.0
	He plasma	19.0	16.5	23.9
	O_2 plasma	24.1	21.6	26.1
Polypropylene	None	2.5	...	...
	He plasma	17.9	...	...
	O_2 plasma	21.2	...	...

employed were single, lap shear, sandwich specimens of 25.4 by 190 mm with a 12.7 mm overlap. The adhesive used was a two-component, polyamide-modified epoxy system comprised of Shell Chemical's Epon 828 epoxy and General Mills' Versamid 140 polyamide. The ratio of Epon to Versamid was 70/30 parts by weight.

Composites

Obtaining maximum performance from advanced fibers, such as Spectra (Allied Signal Inc., Morristown, NJ), is yet another adhesion problem. Unless the fiber bonds well to the resin matrix, stress distribution will be nonuniform and maximum benefit will not be achieved. Spectra, a highly structured polyethylene fiber, provides higher specific strength and stiffness benefits than Kevlar, albeit within a narrower use temperature. But the polyethylene structure has limited property development in structural composites. Plasma treatment makes possible the use of Spectra composites for structural applications as evidenced by the properties shown in Fig. 9.22 for unidirectional composites.

Conclusions

Flame and electric discharges have been used for many decades to introduce oxygen functionality to the surface of plastics to promote adhesion. Cold gas (glow discharge) plasma equipment, suitably rugged for the industrial workplace, has only recently become available. Cold gas plasma allows the user to reengineer the polymer surface, introducing functional chemical groups in a controlled manner. Because the process is conducted in a chamber where the atmosphere and process conditions are precisely controlled, the resulting modification is reliable and repeatable, an attribute very difficult

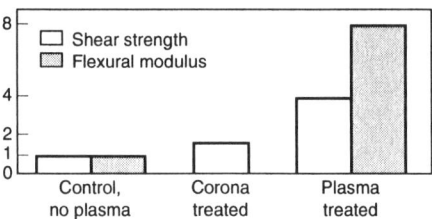

Fig. 9.22 Relative interlaminar shear strength and flexural modulus of epoxy composites as a function of fiber treatment

to ascribe to other surface preparation techniques. The plasma process couples a process gas with an electromagnetic field to create a mixture of active species, such as ions, electrons, and free radicals. Because the process is conducted in a vacuum, it is not in thermodynamic equilibrium; thus, the temperature is not significantly raised, allowing the effective treatment of temperature sensitive materials, such as low density polyethylene moldings and fibers. The radicals, electrons, ions, and photons collide with the polymer, breaking covalent bonds and creating free radicals on the polymer surface. The polymer free radicals react with other species in the plasma environment to establish new surface chemistry. If desired, the plasma process can create high-energy and reactive surfaces essential for maximum adhesion performance.

Importantly, in most applications, the plasma process employs innocuous gases, thereby providing a manufacturing tool that is both workplace and environmentally clean and safe. In addition, the process appears to be universal, with proven effectiveness for all polymer classes (organic and inorganic), albeit the choice of process gas may need to be changed for optimal results.

Industry literature contains numerous references documenting adhesion performance that has been greatly enhanced by plasma treatment. The interested reader will find such documentation to cover the entire spectrum of polymers, from the simplest polyolefin to the most sophisticated of heterocyclic and inorganic polymers. The data presented here are not meant to be comprehensive, but they are rather a representation of the level of improvement typically obtained via plasma treatments. Because of the unusual combination of performance, safety, cleanliness, and cost effectiveness, plasma processing will surely become a commonplace technology in the future.

Chemical

In the past, adhesion promotion through surface preparation accomplished chemically has been somewhat elusive. Plastic materials with low surface energies (polyethylene and polypropylene) have become quite popular in major industrial segments, such as automobile manufacturing, due to their high durability and low cost. Unfortunately, the intrinsic nature of these materials is too unacceptable for painting, coating, or adhesive application.

Certainly the prior discussion regarding corona discharge, flame excitation, and plasma treatment offers processors an opportunity to consider the lower surface energy materials, such as the polyolefins. However, many processors would like to resist the high capital outlay and/or the short shelf life associated with some of these surface activation processes. Chemical methods to activate a lower surface energy plastic surface are desirable. Over the years, these plastics have been treated with primers, tackifiers, even dipped in bleach. None of these techniques has afforded the success seen in other, equipment related, surface excitation processes.

Recent developments with a material called chlorinated polyolefin (CPO) seem to have had success (see Fig. 3.2). CPO is dissolved in a solvent, such as xylene, and applied to the surface of the plastic. The mechanism of how the CPO actually works is not clearly understood, but it is believed that the presence of the chlorine creates more reactive sites to which paints, coating, or adhesives will bond. The CPO may also be integrated directly with compatible paints and coating to increase the adhesion qualities with the requirement of a preparation step.

ACKNOWLEDGMENTS

The author wishes to acknowledge the contributions of Tantec, Inc. to the section "Corona Discharge" and Sherman Treaters to the section "Flame Treatment."

10

Deflashing and Cleaning of Plastic Parts

Unfortunately, plastic parts are often not perfect when they are removed from a mold or die. The nature of the molding operations may result in excess and unwanted plastic material, referred to as "flash," or the parts may have foreign particles or chemicals on their surfaces that must be removed prior to the end-use application. Too often, the removal of flash or the cleaning of the plastic product is not given sufficient attention. This oversight results in the product not meeting the customer's requirements or damage to the plastic part.

Degating

Of the plastic parts molded today, 75% remain on a sprue, runner, or gate when they are removed from the mold (Fig. 10.1). Sometimes it is advantageous to keep the plastic parts on a runner system, such as in a plastic model airplane kit. The runners and gates are used to help organize and identify the individual parts. In the plastic manufacturing environment, this situation would be referred to as a "Kan Ban."

When it is undesirable to have the parts remain on the runner and gates, they must be removed. The most frequent method of part/runner separation is by hand. Whether an operator removes the parts by pulling them off like grapes or using a set of clippers, this repetitive motion is a problem. First, the resulting appearance of the plastic part may be rendered unacceptable,

226 / Decoration and Assembly of Plastics

usually due to excess gate vestige (Fig. 10.2). The clipping action may also result in repetitive motion injuries, such as carpal tunnel syndrome.

Ultrasonic welding equipment can be modified to facilitate the degating of plastic parts from the runners by applying ultrasonic vibration to the

Fig. 10.1 Parts on runners

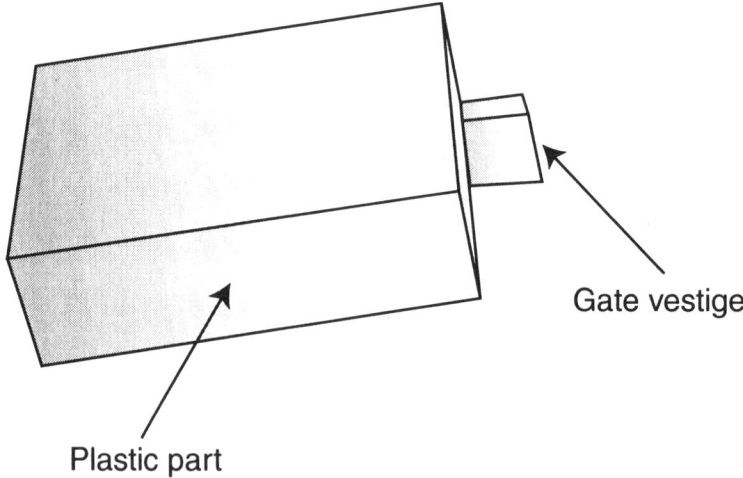

Fig. 10.2 Excess gate vestige

runner. The vibration is translated through the runner system and manifests itself by causing the part to vibrate at a frequency equal to that of the ultrasonic equipment. This vibration results in a flexing of the gate area, which, in turn, causes the gate area to heat up to near the melting point. Next, the weight of the plastic part causes it to detach from the gate (Fig. 10.3).

Flash Removal

Flash is undesirable and nonproductive plastic that results when plastic melt leaks between worn mold details. In some instances, such as the processing of thermosetting plastics, the flash may be small enough to be unnoticeable or not affect the performance of the plastic product in its end-use application. In such an instance, product engineers may add a flash note to the plastic part print, stating that flash is allowable in specified amounts and in specific areas of the part. The flash creates an appearance or visual defect. In any of the above situations, the flash may have to be removed.

The best way to handle a flash problem is not to allow flash to occur in the first place. The way to accomplish a no-flash part is through proper processing and mold maintenance. However, even with due diligence, flash may still occur. There are two types of flash: internal and external.

Internal flash resides in hard-to-reach areas (Fig. 10.4). Internal flash is the most difficult type of flash to remove and usually requires manual deflashing or

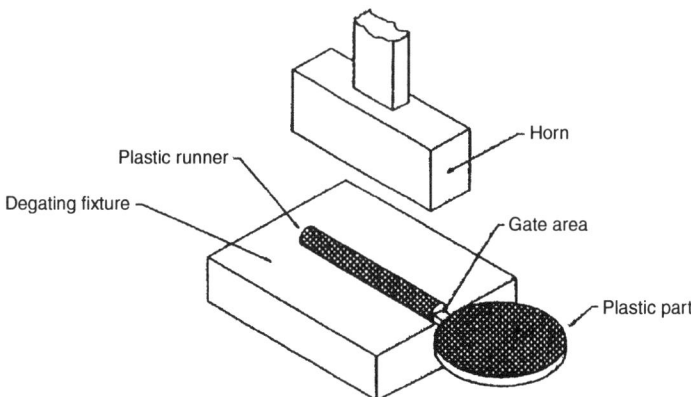

Fig. 10.3 Ultrasonic degating

specialized equipment. It is the most costly to remove because of its location.

External flash resides on the outer edges of the plastic product (Fig. 10.4). External flash may require the same forms of deflashing as internal flash problems, but it is more accessible. Therefore, external flash can be removed more readily than internal flash.

When it is required or desired to remove flash, there are several techniques that can be employed.

Cutting and Trimming

Removing the flash manually is the most popular method, and several specialized flash-removing hand tools have been designed to facilitate the process (Fig. 10.5). It is initially perceived as being the most cost effective, usually when compared against stopping production, removing the mold, and performing repairs. Certainly, manual flash removal offers an additional cost for short-run production scenarios, but it should not become a standard operating procedure. There are several hidden and real costs associated with manual deflashing.

Risk of Operator Injury. Deflashing tools are usually some form of knife, and the potential for injury is always present. Additionally, the repetitive motion associated with manual flash removal could result in other forms of personal injury.

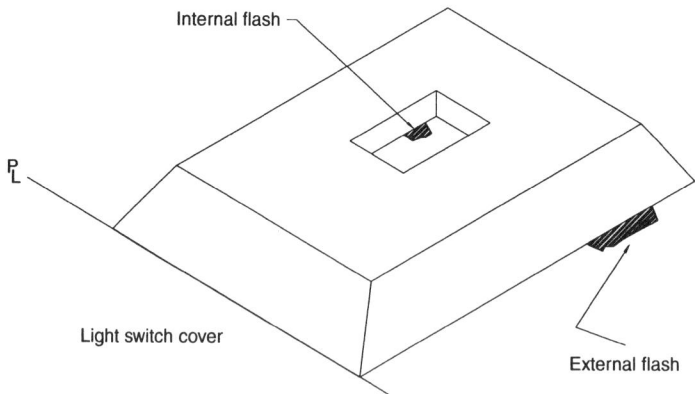

Fig. 10.4 Internal and external flash

Deflashing and Cleaning of Plastic Parts / 229

Risk of Mold Damage. Knives and cutting tools can easily scratch or damage expensive molds. Regardless of how hard the surface of the mold is, a sharp tool can easily scratch or mar the surface.

Damage to the Plastic Part. Flash never fixes itself; it will always increase. Manual flash removal requires a certain operator knack or skill. As the process of removing more and more flash continues throughout the production shift, tired operators will make more mistakes and damage more production parts.

Tumbling

Tumbling is one of the simplest mechanized forms for plastic part deflashing. The equipment is simple (Fig. 10.6) and is often built in-house. The equipment usually consists of a drum fabricated using perforated metal. The drum is nested on motorized rollers and allowed to rotate at speeds as high as 4 cycles/min.

The tumbling process works best with somewhat rigid plastic parts that have a thin brittle external flash. The plastic parts are loaded into the drum and are allowed to tumble much like wet clothes in a clothes dryer. The tumbling action causes the plastic parts to impact each other, thus knocking off the brittle, external flash.

The advantage of tumbling is the low cost. The disadvantages include that it is a batch process requiring constant loading and unloading. The process could damage more fragile plastic parts. Additionally, internal flash can not

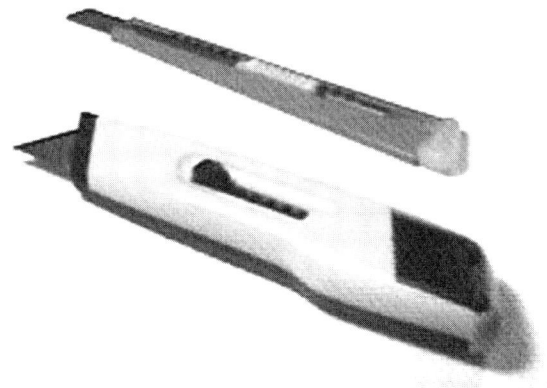

Fig. 10.5 Flash removal tools

be removed if it is inaccessible to the impact of the tumbling parts. Flexible materials, such as elastomers of many unfilled semicrystalline plastics, can not be deflashed using the tumbling process, because the flash on these parts is also flexible and does not break off the parts when impacted.

Media Blasting

Media blasting is a form of plastic part deflashing that picks up where simple tumbling leaves off. Media blasting equipment (Fig. 10.7) utilizes a tumbling apparatus that is incorporated in a closed environment. As the plastic parts are tumbled, an impact medium is broadcast onto the rotating parts under pressure. The media material is a particulate polymer (usually nylon or polycarbonate) or the more classic particulate, walnut shells. This media material will impact the flash causing it to break free of the plastic part. Properly used, media deflashing equipment will not damage the plastic part and will successfully remove both internal and external forms of flash.

The media used in media blasting must be selected so that the blasting material only removes flash and does not destroy the plastic part. Typical materials used for blasting media include walnut shells, apricot pits, polycarbonate, and nylon.

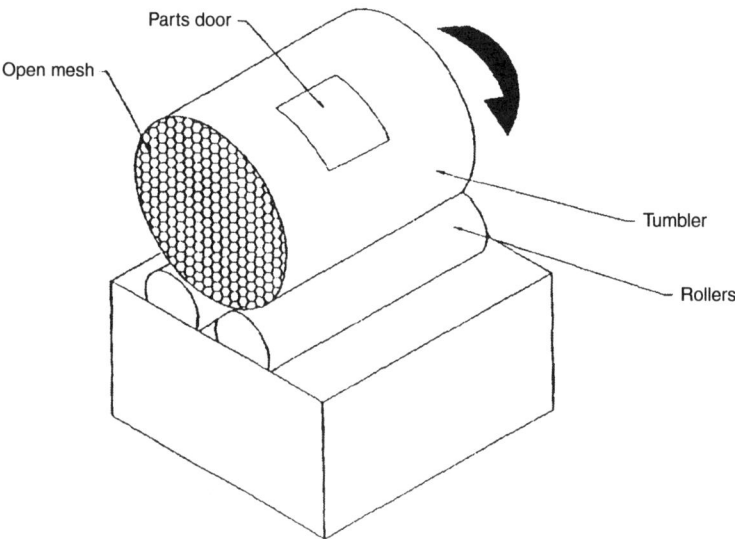

Fig. 10.6 Tumble deflasher

Historically, this form of flash removal has been a batch process, which added to the cost in the operator involvement in the process. New processes are now capable of removing flash from continuously fed product (i.e., move directly from the molding process onto a conveyor system, enter the deflashing equipment, and exit in the correct order), which allows for automation.

Cryogenics

Cryogenics deflashing (Fig. 10.8) utilizes a cold temperature environment in combination with tumbling and/or media blasting to remove internal and external flash from flexible plastic or elastomeric parts. The cold temperature environment is achieved through the use of liquid nitrogen and can achieve temperatures as low as 100 K. Flexible materials and the respective flash are rendered brittle at these low temperatures and are then readily removed with tumbling and media impact. Needless to say, the plastic or elastomeric parts that are deflashed using cryogenics must be sufficiently robust to withstand both the extreme, cold temperature and the impact that occurs at these low temperatures.

Cleaning

Why do some plastic parts need to be cleaned? During the molding process, there are several opportunities for the plastic part to become contaminated with dirt or grease. Poorly maintained equipment can be covered with

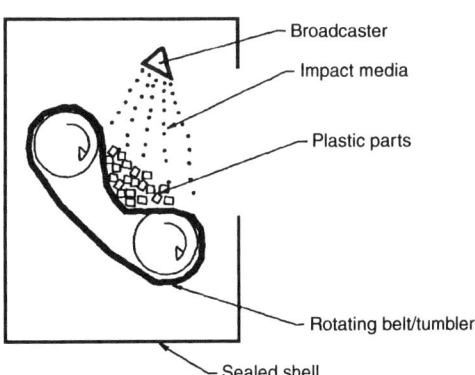

Fig. 10.7 Media deflasher

grease and oil. The simple process of ejecting the plastic part from the mold may result in the part hitting the contaminated portion of the machine. Other contaminants could include common processing aids, such as mold release, oil, and cleaners. One of the most frequent forms of plastic part contaminants is the natural oils that are present when operators touch the plastic parts. This contamination can be worsened when operators use hand creams and moisturizers.

Soaps and Detergents

The basic difference between soap and detergent is that soap, historically, is made from natural materials. A detergent is manufactured utilizing synthetic materials. It is unlikely that a soap would be used to clean plastic parts for the simple reason that soaps tend to leave residues that are often worse than the contaminant they were intended to remove. Detergents can be designed or selected to remove a very specific contaminant without leaving any residue.

Degreasers

Degreasers are usually chemicals designed to remove contaminants, such as grease, oil, and particulates. This cleaning operation can be accomplished by the simple action of spraying a chemical solvent onto the contaminated plastic part or using a more sophisticated equipment-based system referred to as ultrasonic degreasing.

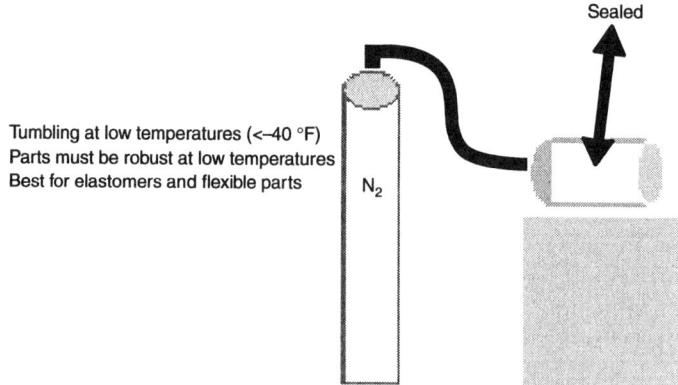

Fig. 10.8 Cryogenic deflasher

Deflashing and Cleaning of Plastic Parts / 233

Ultrasonic degreasers (Fig. 10.9) utilize some of the same basic technology described in the section on ultrasonic welding (see the section "Degating"). The basic concept involves a chemical tank that holds the desired chemical solvent. Beneath the chemical tank is an ultrasonic generation mechanism that induces microscopic bubbles throughout the chemical in the tank. The movement of these bubbles, along with the chemical itself, results in a very thorough cleaning action.

One obvious point to consider is the compatibility of the chemical solvent and the plastic being cleaned. There have been, unfortunately, many instances where the process engineer did not consider the chemical compatibility of the plastic with the degreasing media, and this oversight resulted in either a degradation of the plastic part or the disappearance of the plastic

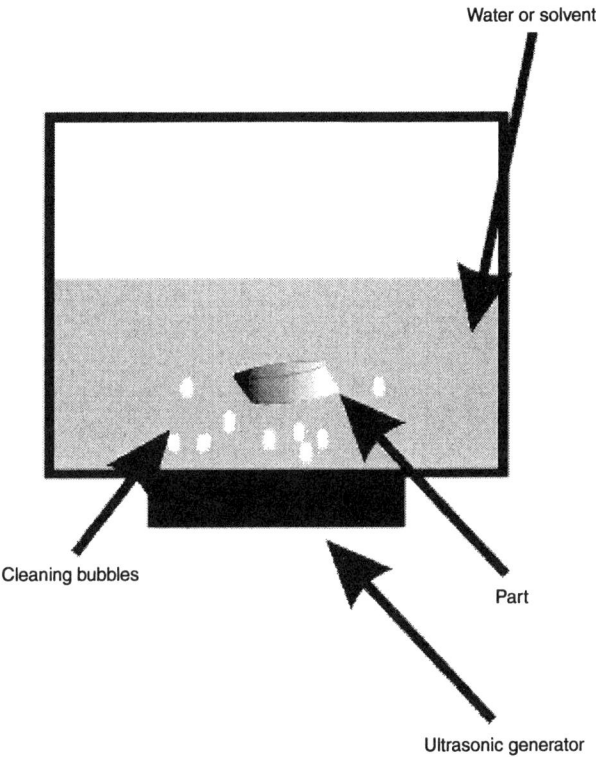

Fig. 10.9 Ultrasonic degreaser

part completely. Typical solvents used in degreasing systems include water, Freon-based (E.I. Du Pont de Nemours & Co., Inc., Wilmington, DE) materials, and chlorinated hydrocarbons. Many of the more effective chemical solvents are either banned or regulated by various government agencies. It is imperative that all aspects of environmental regulations be investigated prior to using chemical solvents and degreasing systems.

Activated Gas Plasma Cleaning

A plasma is a partially ionized gas containing electrons, ions, and various neutral species at many levels of excitation (Fig. 10.10). Ionization of the gas molecules is accomplished by subjecting the gas, which is enclosed in a vacuum chamber, to radio frequency (RF) energy. Gases or mixtures of gases used for cleaning and contaminant removal include argon, nitrogen, oxygen, air, tetrafluoromethane, and helium. The radicals and active species generated in the plasma react with the surface contaminants to produce volatile byproducts that include water vapor, carbon monoxide, and carbon

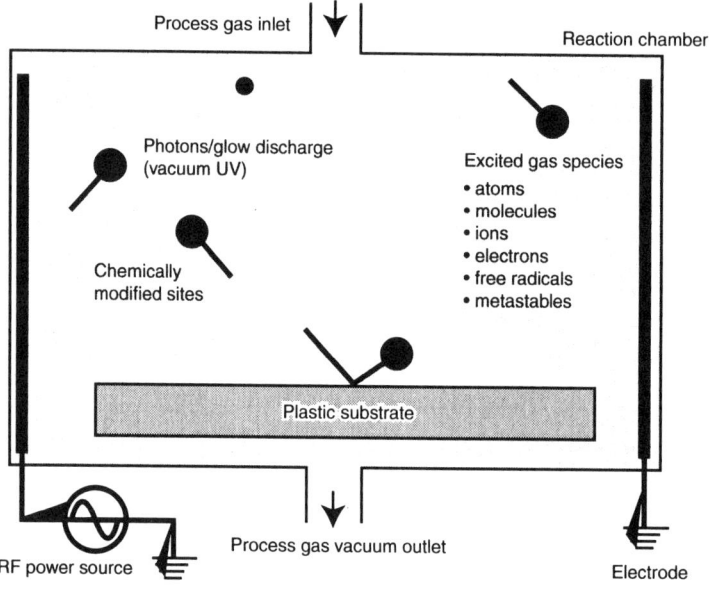

Fig. 10.10 Plasma cleaning

dioxide. The nature of the volatile products created is dependent on the gas chemistry used for the cleaning process. These reaction products are minimal and are safely pumped from the reaction chamber by the vacuum system.

When the substrate to be cleaned consists of oxidizable materials, the concentration of oxygen in the plasma feed gas and the process time must be properly controlled. In such situations, inert gases, such as argon, helium, and nitrogen, have been found to be effective in removing contaminants. In addition to the choice of gas chemistry to be used, the variables in a plasma cleaning process are the vacuum level, gas flow rate, the energy applied to ionize the gas, and the process time in the plasma.

Although the electron temperature in a plasma can be as high as 5000 K, the bulk temperature of the gas in the reactor is essentially at ambient. Thus, activated gas plasmas are also being used to remove contamination from plastic surfaces. Indeed, this removal is the first step that typically occurs in the process of surface modification of plastic materials using a cold gas plasma. Through the use of appropriate process parameters, highly effective cleaning can be achieved without adversely affecting the underlying substrate.

Environmental Effects

According to one of the leading manufacturers of plasma systems for the plastics industry (Plasma Science), Ecoserve Environmental Services (Pittsburgh, CA), an independent laboratory and licensed air testing organization, was contracted to test stack emissions at Plasma Science's laboratories in California. Testing was conducted in accordance with the Environmental Protection Agency's (EPA) code of Federal Regulations (Title 40, Part 60, Appendix A, Methods 3 and 4, 1985; Appendix A, Method 5, 1981). The NO_X, CO, and SO_2 were continuously monitored during this test. The CO and SO_2 remained at ambient levels during the entire 6 h test. The NO_X was measured at 6 ppm when N_2O was being used in the process (not continuously in use). Oxygen was also analyzed during this test and found to be at ambient levels. Particulate emissions were measured at a level of 0.001 lb/h. At this rate, an increase of orders of magnitude in particulate emission rate could be tolerated without exceeding government mandated or recommended limits.

All of the case studies cited suggest the use of benign and innocuous gases for plasma cleaning processes. Under these circumstances, environmental damage is avoided, while superior process performance is achieved.

Studies have shown plasma-based processes to be very effective in removing surface contamination from different materials. Analysis of exhaust gases from plasma reactors has verified ambient levels of chemicals and extremely low particulate matter. The effectiveness of innocuous gases in achieving contamination removal further strengthens the role of plasma processes as the environmentally safe alternative to chemical methods of contaminant removal.

INDEX

A

Ablation 217
ABS. *See* Acrylonitrile butadiene styrene.
ABS/PC. *See* Acrylonitrile butadiene styrene/polycarbonate.
ABS/PVC. *See* Acrylonitrile butadiene styrene/polyvinyl chloride.
ABS/SAN. *See* Acrylonitrile butadiene styrene/styrene-acrylonitrile.
Acetal
   compatibility with other thermoplastics .... 58(F)
   ultrasonic welding 63
Acetal copolymer
   ultrasonic welding characteristics .... 57(T)
Acetal homopolymer
   ultrasonic welding characteristics .... 57(T)
Acoustic impedance
   of horn material 77
   specific (Z_1) 77
Acoustic impedance characteristic (Z) 77
Acrylic adhesives, corona discharge 202
Acrylic multi polymer
   compatibility with other thermoplastics .... 58(F)
Acrylic multipolymer XT
   ultrasonic welding characteristics .... 57(T)
Acrylic paint systems 185
Acrylic/polyvinyl chloride
   ultrasonic welding characteristics .... 57(T)

Acrylics
   amorphous structure 16
   colorant use in 20
   compatibility with other thermoplastics .... 58(F)
   as conformal coating 189
   cumulative domestic growth .... 16(F)
   electrolytic plating 171
   impact modified, ultrasonic welding characteristics .... 57(T)
   in ABS copolymer 17
   internal coloring for 181
   surface free energy .... 208(F)
   ultrasonic welding .... 57(T), 63, 64
   vacuum metallization 163
Acrylite, electrolytic plating 171
Acrylonitrile
   electrolytic plating 171
   vacuum metallization 163
Acrylonitrile butadiene styrene (ABS)
   application, induction inserting .... 94, 95(F)
   colorant use in 20
   compatibility with other thermoplastics .... 58(F)
   copolymer structure .... 17–18(F)
   electrolytic plating 171, 172
   internal coloring for 181
   ultrasonic welding .... 57(T), 63, 64
   vacuum metallization 163
   with proportions of material components varied ($_{AB}$S, ABS, AB$_S$) 171–172
Acrylonitrile butadiene styrene/ polycarbonate (ABS/PC) 17(F), 18

Acrylonitrile butadiene styrene/ polycarbonate (continued)
compatibility with other thermoplastics .... 58(F)
ultrasonic welding characteristics .... 57(T)
Acrylonitrile butadiene styrene/ polyvinyl chloride (ABS/PVC)
compatibility with other thermoplastics .... 58(F)
ultrasonic welding characteristics 57(T)
Acrylonitrile butadiene styrene/ styrene-acrylonitrile (ABS/SAN)
cumulative domestic growth .... 16(F)
Acrylonitrile-styrene-acrylate terpolymer (ASA)
ultrasonic welding characteristics .... 57(T)
Activated gas plasma cleaning .... 236–238(F)
environmental regulations 237–238
Activation 217, 218
Additives .... 12, 18–25(F)
in plastic compounds 18
Adherend, defined 30
Adhesion .... 30–31(F)
Adhesives .... 29–43(F)
bonding mechanisms .... 31–33(F)
chemical families 37
classification scheme 35–37
consumer 33, 35
defined 29
factors affecting bond strength .... 40–43(F)
film format 40
functions 33
function type 35–37
hot melt 37, 40
hot melt, as chemical family 37
industrial 33, 35
loading environment 41–42
material selection matrix ...... 34–35(F), 36(F)
multicomponent systems 40
natural polymer 37
one-component systems 40
physical forms 40
pressure sensitive (PSA) 37
property assessment .... 40–43(F)
selection process 33–34
setting or curing temperature range 36
structural 36
substrate loading scenarios ... 41–42(F)
synthetic polymer 37
terminology .... 29–31(F)
thermoplastic .... 37, 39(F)
thermoset .... 37, 38(F), 39(F)
two-component system .. 29, 30(F), 40
types 33
Air-blown arc plasma treaters 202–203
Air/butane mixture ratio 210
Air/methane mixture ratio 210
Air/propane mixture ratio 210
Alloy 18
Aluminum
as fixture material for pad transfer printing 155–157
as fixturing component material for vertical stamping machines 123
induction inserting, metal to plastic bonding .... 86(T), 87(T), 91
as silicone rubber transfer pad base material 153
as ultrasonic assembly horn material 77, 78
Aluminum stearate 55
American Society of Electroplated Plastics (ASEP)
basic service conditions for electroplated plastics 178
American Society for Testing and Materials (ASTM)
vacuum metallization tests 169
Ameritherm Inc. 83
Amides 204
plasma surface treatment 220
Amination
to increase surface energy 217
Ammonia
in plasma surface treatment 217
Aniline, surface tension .... 208(F)
Annular snap 7
Antiblock agents 204
Antistatic agents .... 18, 22–23(F)
categories .... 23(F)
external 23

internal 23
ion discharge .... 23(F)
Arc discharge 202
Arcing
   from corona discharge systems 215
Argon, in plasma surface treatment ..217
Aromatic esters
   plasma surface treatment 220
ASA. *See* Acrylonitrile-styrene-acrylate terpolymer.
Assembly of plastic parts, equipment 5
Assembly operation .... 1, 4(F)
Autocatalytic process 174
Automobile assembly
   applications of heat staking 100

B

BDS. *See* Butadiene styrene.
Beading 26, 27
Benzene .... 11(F)
   surface tension .... 208(F)
Biocides 25
Blending .... 17–18(F)
Blow molding 1
Bonding 25
   mechanisms .... 31–33(F)
   terminology use 30
Branched polymers 18
Branching 215
Brass
   as die material for hot stamping....124, 125(F), 130(T), 134
   induction inserting, metal to plastic bonding .... 86(T), 87(T), 89(F), 91
   threaded inserts for induction inserting .... 79–80, 81–84(F,T)
Brass inserts
   large, induction inserting of 97–98(F)
Brushing 189
Burner ignition method
   gas flame treatment 214
Butadiene
   electrolytic plating 171
   in ABS copolymer 17
Butadiene styrene (BDS)
   compatibility with other thermoplastics .... 58(F)
   ultrasonic welding characteristics .... 57(T)

C

CA, CAB, CAP (cellulosics)
   ultrasonic welding characteristics 57(T)
Carbon bisulfide
   surface tension .... 208(F)
Carbon black
   as black colorant 21
   effect on rigidity of plastic 21
   UV stability as secondary benefit 22
Carbon chain, in plastics .... 10(F)
Carbon/graphite fibers
   as reinforcement materials 19–20
Carbonyl group 201
Carboxyl group.................................. 201
Casting compounds
   as fixture material for pad transfer printing 155–157
Cellulose .... 11(F)
Celulosics
   compatibility with other thermoplastics .... 58(F)
   feedstock materials .... 11(F)
   surface free energy..................... 208(F)
   ultrasonic welding characteristics .... 57(T)
Ceramics, surface free energy ..... 208(F)
Chair frame
   as application of induction inserting .... 92–93, 94(F)
Chemical absorption .... 31–32(F)
Chemical coupling agent
   for reinforcement materials 19
Chemical (covalent) bonds 32
Chemical surface treatments ..223–224
Chlorinated polyolefin (CPO) ...... 31(F)
   chemical surface treatment 224
Chromate dips 165
Chromium, hot stamping foils 129
Cissing 26, 27
Clean Air Act of 1990 6

Cleaning .... 233–236(F)
Cleavage, bond strength .... 42(F)
Cliché .... 143(F), 145(T), 147(F), 148, 149
   etch depth and applications ...... 149(T)
Cloth, as filler material 20
Coating
   gas flame treatment level requirements 207
   gas flame treatment theory 209
Colloidal suspension 174
Colorants 18, 20, 181
   advantages and disadvantages 21
   forms and range added 20
   pigments 182–183
   spin welding compatibility factor 46
Coloring matching booth 182
Compact disc (CD) .... 6(F), 7
Compatibility 18
Compression
   silicone rubber formulations .... 128(T)
Conformal coatings .... 188–189(F)
Contact angle .... 26(F), 27
   vs. surface energy measurements ... 219
Control arm/handle assembly
   as application of induction inserting .. 96
Copolymerization .... 17–18(F)
Copolymers .... 17(F)
   ultrasonic welding 64
Copper
   as die material for hot stamping .. 124, 125(F), 130(T), 134
   as ultrasonic assembly horn material 77
Copper-accelerated acetic acid-salt spray (CASS) tests
   vacuum metallized parts 169
Corona discharge .... 199–204(F), 205(F), 206(F)
   equipment .... 202–203(F)
   shelf life of treated surfaces 204
   surface cleanliness 204
   surface modification with electrical discharges .... 200–202(F)
Cost model, basic .... 1, 4(F)
   injection molded part 1–2
Cotton, as filler material 20
Covalent bonds 32

CPO. See Chlorinated polyolefin.
Crazing 183
Crosslinking .... 12–13(F), 14(F), 215, 217, 218
Cryogenics deflashing .... 233, 234(F)
Curing 37
   multicomponent thermoset adhesives 40
Cyanoacrylates 37
   2-cyanoacrylate 40, 41
Cyclohexane, surface tension ...... 208(F)

D

Date code .... 1, 2(F)
Decoration and assembly standard cost 2
Decoration of plastic parts
   equipment for 5
Decoration operation .... 1, 4(F)
Deflashing and cleaning
   of plastic parts .... 227–238(F)
Deflection temperatures
   reinforcement effect 18–19
Degating .... 227–229(F)
   ultrasonic .... 228–229(F)
Degreasers .... 234–236(F)
Degree of polymerization 2
Delrin 503
   plasma surface treatment .... 221(T)
Desiccants 55
Detergents 234
Dielectric sealing .... 78–79(F)
Diffusion model .... 31, 32(F)
Dipping 189, 192–193
Doctor blade .... 143(F), 145(T), 148–149, 150(F)
   configurations .... 150(F)
   life 150
Drums, cardboard .... 12, 13(F)
Durel
   plasma surface treatment .... 221(T)
Dwell time .... 118–119(F)
Dyes, for in-house coloring 182

E

Ecoserve Environmental Services ... 237
E-glass 19
Electrical connectors
   as application of induction
      inserting .... 95–96, 97(F)
Electroless plating
   in preplate process .... 173–174(F)
Electrolytic plating
   acid copper bath 175
   applications 179
   bright nickel bath .... 176(F), 177
   chrome bath .... 176(F), 177
   copper strike .... 175, 176(F)
   design accommodations .... 179–180(F)
   etching .... 172(F)
   microporous nickel-Dur-Ni bath 177
   preplate 172
   quality measurement 178–179
   racking plastic parts .... 177–178(F)
   selection of plastic material ... 171–172
   semibright nickel bath .... 176(F), 177
   washing and neutralizing 177
Electromagnetic interference
   (EMI) reduction 161
Electromagnetic interference
   (EMI)shielding
   as application of heat staking 100
Electron spectroscopy for chemical
   analysis (ESCA) 218, 220
Electroplating .... 1, 3(F)
Electrostatic attraction 32–33
Electrostatic spray systems 190–192(F)
Elongation
   silicone rubber formulations .... 128(T)
Enamels 185
Encapsulated heat stations 83
Energy directors .... 62–64(F), 65(F)
Energy required to heat
   metal insert 87
Engineered thermoplastic
   materials .... 17–25(F)
Engineering plastics
   feedstock materials .... 11(F)
   plasma surface treatment 220–221
Engineering thermoplastics

cumulative domestic growth .... 16(F)
Engraved plate .... 148–149(F,T)
Engraved plate technology angled
   image technique .... 149(F)
Engraved plate technology screened
   image technique .... 149(F)
Environmental emission standards 6
Environmental regulations
   disposal of finished parts 3
EPDM. See Ethylene-propylene-diene
   monomer.
Epon 828 epoxy
   (Shell Chemical) .... 221(T), 222
Epoxies 37, 40
   as conformal coating 189
   cumulative domestic growth .... 14(F)
   fillers present 33
   in top coat systems for vacuum
      metallization 167
   structural formula .... 38(F)
Epoxy adhesives, corona discharge 202
Epoxy composites
   plasma surface treatment vs.
      Corona treatment .... 222(F)
Epoxy paint systems 184
Equipment
   for assembly 5
   for decoration .... 5(F)
Erosion
   of mold and molding machinery 19
Etch baths 172
Etch depth .... 149(T)
Etching, in preplate process .... 172(F)
Ethane .... 11(F)
Ethyl alcohol, surface tension ..... 208(F)
Ethylene-propylene-diene monomer
   (EPDM) rubber
   corona discharge 203
External systems 24
Extrusion 1

F

Far-field welding 55
Feedstock materials ..... 11–12(F), 13(F)
   for plastics .... 11(F)

FEP. *See* Fluorinated ethylene-propylene.
Fiber orientation .... 19(F)
Fillers 18, 20
   advantages and disadvantages 20
   disadvantages 20
   and ultrasonic welding 56
Fin and knurl design .... 81–82(F)
Flame retardants 18, 21
   advantages and disadvantages 21
   categories 21
   and ultrasonic weldability 56
Flame (gas) treatment .... 204–214(F)
Flash 82, 227
   cutting and trimming .... 230–231(F)
   external ..... 229–230(F), 231, 232, 233
   internal ...... 229–230(F), 231–232, 233
Flash-off 185
Flash removal 229–233
   cryogenics deflashing .... 233, 234(F)
   hazards 230–231
   media blasting .... 232–233(F)
   tools for .... 230–231(F)
   tumbling 231–232
Flash traps
   spin welded joints .... 47, 49(F)
   vibration welded joints .... 51(F)
Flash wells
   scarf joints .... 69(F)
   shear joints .... 66, 67(F)
Flexo printing
   gas flame treatment level
      requirements 207
Flexural modulus
   reinforcement effect 18–19
Flood bar 143
Flow coating 192–193
Flow rates 47
Fluorinated ethylene-propylene (FEP)
   peel strength after plasma surface
      treatment .... 220(T)
Fluorine-containing plasmas 218
Fluorine free radicals, oxidation by 217
Fluorocarbons 55
Fluoropolymers
   peel strength after plasma surface
      treatment .... 220(T)
   plasma surface treatment 219–220

ultrasonic welding
   characteristics .... 57(T)
Flush stamping .... 123–124(F)
Foaming agents 18, 24
   foaming techniques 24
Foils .... 129–139(F,T)
   cradle and mandrel type part holding
      fixtures .... 134, 135(F)
   day-glo pigment foil
      construction .... 132, 133(F)
   diffraction 129–130
   flat hot stamping dies used 134
   gloss pigment foil
      construction .... 131–132(F)
   gold metallic ..... 133, 134–136(F), 139
   holographic 3-D 129–130
   hot stamping .... 115–119(F),
                            120(F), 121, 124
   matte pigment foil
      construction .... 131, 132(F)
   metallic foil construction   130–131(F)
   moving head method of roll-on
      decorating .... 136, 137(F)
   peripheral marking
      technology .... 132–136(F)
   pigment foil construction 131–132(F),
                                  133(F)
   reciprocating table part transfer
      technique for roll-on
      decorating .... 137, 138(F)
   roll-on decorating
      technology .... 136–139(F,T)
   silver metallic ... 133, 134–136(F), 139
   types of colors and designs 129
   wood grain designs 136, 139
Formamide
   surface tension .... 208(F)
Free radicals 218
   on polymer surface 201
   plasma surface treatment 223
Fusion bonding .... 47–50(F)
   advantages and disadvantages ... 48–49
   applications 48
   equipment .... 47–48, 50(F)
   equipment costs 48
   joint area design 50
   material compatibility 48

G

Gas flame treatment .... 204–214(F)
   benefits 206–207
   electronic control 214
   flame geometry .... 211–214(F)
   gas/air mixture
      control .... 210–211(F), 212(F)
   theory .... 209–210(F)
   treatment level requirements 207
Gas plasma cleaning .... 236–238(F)
Gates .... 227–228(F)
Gaylord containers .... 12, 13(F)
Glass
   effect on ultrasonic weldability 56
   as reinforcement material .... 19(F)
   surface free energy .... 208(F)
Glass–filled materials
   internal coloring for 181
Glow discharge 202
Glue, defined 29
Gluing
   gas flame treatment level
      requirements 207
   terminology use 29
Glycerol, surface tension .... 208(F)
Glycol, surface tension .... 208(F)
Graft polymers 18
Granulation 2
Graphics
   multicolored 115
   permanent gold and silver metallic 115
Gravelometer 169
Gravure printing
   gas flame treatment level
      requirements 207

H

Half-wave length
   of simple cylindrical horn 76
Hand-held colorimeter 182
HDPE. See High-density polyethylene.
Heat resistance
   silicone rubber formulations .... 128(T)
Heat staking .... 100–110(F,T)
   advantages and disadvantages 100,
                      104–105, 106(F)
   applications 100
   description 100
   direct contact (heated probe)
      method .... 100, 101(F),
                   102–104(T), 109(F)
   disassembly of stakes .... 109(F)
   hot air/cold staking .... 100, 101(F),
               102,103(T), 106(F), 108, 109
   hot gas welding .... 110–111(F)
   methods .... 100, 101–104(F,T)
   post clearance hole in the
      attachment .... 108(F)
   post features .... 105, 106(F), 107(F)
   probe clearance .... 107(F), 108
   reassembly 109
   side swaging .... 109–110(F)
   stake head .... 105–107(F)
   strength of a stake, calculation of 108
   ultrasonic staking 100–102,
                        103(T), 105
Heat transfers, preprinted 120
Helium
   in plasma surface treatment 217
High-density polyethylene (HDPE)
   cumulative domestic growth .... 16(F)
   plasma surface treatment .... 221(T)
   surface free energy .... 208(F)
   ultrasonic welding
      characteristics .... 57(T)
High-frequency arc treaters
   (EST systems) .... 202(F), 203
Holograms
   security labels for compact discs 7
Homopolymers, ultrasonic welding 64
Horn tuning 78
Horn velocity 77–78
Horn velocity amplitude (V) 77
Hot plate welding .... 47–50(F)
Hot stamping .... 6, 113–139(F,T)
   advantages .... 113–116(F), 117(F)
   applications .... 113, 114(F),
                  115(F), 116(F), 117(F)
   foils .... 129–139(F,T)
   graphics on computer discs 7
   manufacturing code on
      compact discs 7

Hot stamping (continued)
  permanent gold and silver metallic
    graphics produced 115
  process .... 116–119(F)
  troubleshooting guide .... 139(T)
  vertical hot stamping machine
    selection guide .... 121, 122(F)
  vertical stamping
    technology .... 120–128(F,T)
Hydrolysis
  to increase surface energy 217
Hydroperoxide group 201
Hydroxyl group 201

I

Ianomer (high)
  surface free energy .... 208(F)
Ianomer (low)
  surface free energy .... 208(F)
Imides, plasma surface treatment 220
Indexing conveyors 122, 147
Induction bonding .... 99–100(F)
  advantages 100
Induction inserting .... 79–98(F,T)
  applications .... 92–98(F)
  inserting metal into
    plastic .... 81–84(F), 85(F)
  inserting metal parts
    accurately .... 92, 93(F)
  metal threaded inserts 79–80
  metal to plastic bonding ..... 84–91(F,T)
    effects on properties 85, 86–87(T)
    material insertion process... 85(F),
      87–91(F)
  multiposition coil 91–92
  parameters for consistent
    process .... 82(T)
  simultaneous insertion of three
    steel inserts 91–92
In-house coloring 182–183
Injection molding 1
  cost model 1–2
  jewel case for compact disc 7
Ink jet printing 197
Inks, for pad transfer printing... 153–155

Interference snap, subtle 7
Internal blowing agents 24
Internal coloring 181–183
Isopropyl alcohol
  surface tension .... 208(F)

J

Joe Blocks .... 31, 32(F)
Joint designs .... 62–69(F,T)
  butt joint with an energy
    director .... 62–64(F)
  interference .... 66(F), 67(F,T), 68(F)
  scarf .... 68–69(F)
  shear .... 66(F), 67(F,T), 68(F)
  step .... 64(F), 65(F)
  tongue and groove .... 64, 65(F)

K

"Kan Ban" 227
Kaolin (clay), as filler material 20
Knurl and fin design .... 81–82(F)

L

Labor (L) .... 1–2, 4(F)
Laminating
  gas flame treatment theory 209
  solvent-based, gas flame treatment
    level requirements 207
  water-based, gas flame treatment
    level requirements 207
Landfill regulations
  disposal of finished parts 3, 6
Lap joints, adhesive-bonded ....42–43(F)
Laser etching
  manufacturing code on
    compact discs 7
Laser printing/etching .... 193, 194(F)
Latex paint systems 185
LDPE. *See* Low-density polyethylene.
LDPE/LLDPE. *See* Low-density
  polyethylene/LLDPE.

Lead
as ultrasonic assembly horn
material 77
Lead zirconate titanate (PZT) 73–74
Legal regulations 6
Lexan 121
plasma surface treatment .... 221(T)
Liquid crystal polymers
plasma surface treatment 220
Litho/offset printing
gas flame treatment level
requirements 207
Load analysis 42
Low-density polyethylene (LDPE)
characteristics 15
hot melt adhesives 37
plasma surface treatment 223
surface free energy .... 208(F)
ultrasonic welding
characteristics .... 57(T)
Low-density polyethylene/LLDPE (LDPE/LLDPE)
cumulative domestic growth .... 16(F)
Lubricants, and ultrasonic welding 56
Lucite, electrolytic plating 171

M

Magnesium
as die material for hot
stamping ... 124, 125(F), 130(T), 134
Magnification ratio
of booster horn .... 74, 75(F)
Makrozwitterion
structural formula .... 39(F)
Manufacturing code 7
Material cost (M) .... 1–2, 4(F)
Material cost component 2
Materials classification .... 12–16(F)
Material selection
matrix .... 34–35(F), 36(F), 42
Melting point 15–16
Melting range 15–16
spin welding compatibility factor 46
Melt viscosity 10
Mers 9

Metal control arms
as application of induction
inserting 96
Metallization of plastics ...... 161–180(F)
applications .... 161, 162(F)
categorization of applications 161
electroless plating .... 170–180(F)
electrolytic plating .... 170–180(F)
vacuum metallization .... 161–170(F)
Metals, surface free energy .... 208(F)
Methane .... 11(F)
Methyl alcohol, surface tension .. 208(F)
Mica 20
Microballoons 24
Microspheres 24
Modification, of polymers .... 17–18(F)
Modified phenylene oxide
ultrasonic welding
characteristics .... 57(T)
Molded-in bosses .... 47, 48(F)
Mold release agents 165
and ultrasonic welding 55–56
Monomer, defined 11
Moving (x-y) bed 193
MPPO
compatibility with other
thermoplastics .... 58(F)
Multicomponent paint
systems .... 184–185(F)
Multiposition coil .... 84, 85(F)

N

NAS
ultrasonic welding
characteristics .... 57(T)
Natural gas .... 11(F)
n-Butanol, surface tension .... 208(F)
Near-field welding 55
Nesting hinge, simple 7
n-Hexane, surface tension .... 208(F)
Nitration, to increase surface energy 217
Nitrogen
in plasma surface treatment 217
Nitrous oxide
in plasma surface treatment 217

Noryl, vacuum metallization 163
Noryl 731
   plasma surface treatment .... 221(T)
Nylon
   colorant use in 20
   compatibility with other
      thermoplastics .... 58(F)
   as fixture material for pad transfer
      printing 155–157
   as fixturing component material for
      vertical stamping machines 123
   impact medium for media blasting 232
   induction bonding 100
   polyamide, hot melt adhesives 37
   semicrystalline structure 16
   surface free energy .... 208(F)
   ultrasonic welding .... 55, 57(T), 63
   vacuum metallization 163
Nylon 6
   plasma surface treatment .... 221(T)

O

Oil .... 11(F)
Overhead (O) .... 1–2, 4(F)
Overlap joints, adhesive-bonded .. 43(F)
Oxygen, in plasma surface treatment 217
Oxygen analyzer and feedback
      system .... 212(F)
Oxygen plasma, cold gas 216–217
Ozone
   generated by corona discharge
      treatment 215

P

Packaging systems .... 12, 13(F)
Pad transfer printing ...... 141–159(F,T)
   advantages .... 141–142(F)
   applications .... 142(F), 147
   doctor blade technology .... 150(F)
   engraved plate
      technology .... 148–149(F,T)
   four-color process 142
   graphics on computer disc 7
   ink monitoring pump 157, 158–159(F)
   ink technology 153–155
   methods used .... 145, 146(F)
   part-holding fixture
      technology .... 155–157(F)
   process .... 143–145(F,T)
   sealed ink reservoir .... 157–158(F)
   sealed ink reservoir pad printing
      machine .... 145, 146(F)
   transfer pad technology .... 151–153(F)
   troubleshooting guide .... 145(T)
   vertical printing
      technology .... 145–147(F)
   viscosity control
      technologies .... 157–159(F)
   wet-on-wet capability 141
Painting .... 183–188(F)
   application techniques ...... 189–193(F)
   dipping 189, 192–193
   electrostatic spray systems 190–192(F)
   flow coating 192–193
   gas flame treatment theory 209
   paint system components 187–188(F)
   powder coats 186
   properties of the substrate 183
   soft paints .... 186–187(F)
   solvent-based systems 185–186
   spray painting basics, air
      and airless .... 190, 191(F)
   to match coloring 182
   types of paints .... 183–187(F)
Palladium chloride, as catalyst ... 174(F)
Pantone color numbers 153–154
PAR. *See* Polyarylate.
PC. *See* Polycarbonate.
PCB. *See* Printed circuit board.
PC/PET. *See* Polycarbonate/
   polyethylene terephthalate.
PE. *See* Polyethylene.
PEEK. *See* Polyetheretherketone.
Peel, bond strength .... 42(F)
Perfluoro alkoxy (PFA)
   peel strength after plasma surface
      treatment .... 220(T)
Permeability
   induction inserting .... 85, 86(T), 87(T)
PET. *See* Polyethylene terephthalate.
Petroleum .... 11(F)

PFA. *See* Perfluoro alkoxy.
Phenol, surface tension .... 208(F)
Phenolics
   cumulative domestic growth .... 14(F)
   vacuum metallization 163
Phosphate dips 165
Photoengraving, in-house 148
Piezoelectric converters .... 73, 74(F)
Pigments
   for in-house coloring 182–183
   and ultrasonic weldability 56
Plasma cleaning .... 236–238(F)
Plasma-induced oxidation
   to increase surface energy 217
Plasma polymerization 167
Plasmas 201
Plasma surface treatment 215–223(F,T)
   commodity resins .... 221–222(T)
   composites .... 222(F)
   conclusions 222–223
   effectiveness of plasma
      processes 219–220
   engineering plastics 220–221
Plastic bearings onto steel shafts
   as application of induction
      inserting .... 95, 96(F)
Plastic compound 18
   defined 9, 18
Plasticizers 18, 22
   extraction 22
   and ultrasonic welding 56
Plastics
   polar 78–79
   pricing .... 11, 12(F)
Polar functional groups 201–202
Polyacetals
   plasma surface treatment 220
Polyacrylonitrile butadiene styrene.
   See Acrylonitrile butadiene styrene.
Polyamide. *See also* Nylon.
   hot melt adhesives 37
Polyarylate (PAR)
   compatibility with other
      thermoplastics .... 58(F)
   ultrasonic welding
      characteristics .... 57(T)
Polycarbonate (PC)
   amorphous structure 16
   blended with ABS .... 17(F), 18
   colorant use in 20
   compatibility with other
      thermoplastics .... 58(F)
   cost 18
   impact medium for media blasting 232
   internal coloring for 181
   plasma surface treatment 220
   ultrasonic welding .... 55, 57(T), 63
   ultraviolet stabilization 22
   vacuum metallization 163
Polycarbonate/polyethylene
   terephthalate (PC/PET)
   compatibility with other
      thermoplastics .... 58(F)
Polyester
   cumulative domestic growth .... 14(F)
   as foil material for hot stamping ... 129
   thermoplastic, semicrystalline
      structure 16
   thermoplastic, ultrasonic welding 63
Polyester (Mylar)
   surface free energy .... 208(F)
Polyester -PBT
   compatibility with other
      thermoplastics .... 58(F)
   ultrasonic welding
      characteristics .... 57(T)
Polyester-PET
   compatibility with other
      thermoplastics .... 58(F)
   ultrasonic welding
      characteristics .... 57(T)
Polyetheretherketone (PEEK)
   compatibility with other
      thermoplastics .... 58(F)
   ultrasonic welding
      characteristics .... 57(T)
Polyetherimide
   compatibility with other
      thermoplastics .... 58(F)
   ultrasonic welding
      characteristics .... 57(T)
Polyethylene (PE)
   adhesion to 31
   compatibility with other
      thermoplastics .... 58(F)
   corona discharge 199

Polyethylene (continued)
  feedstock materials .... 11(F)
  internal coloring for 181
  not metallized 163
  pad transfer printing 154
  plasma surface treatment 222
  semicrystalline structure 16
  surface characteristics 25
  surface energy 200
  ultrasonic welding ....57(T), 63
Polyethylene terephthalate (PET)
  cumulative domestic growth .... 16(F)
  ultrasonic welding
    characteristics .... 57(T)
Polymer
  defined ....9(F)
  feedstock materials .... 11(F)
  modification of .... 17–18(F)
Polymer blends .... 17(F), 18
Polymer fabrics
  feedstock materials .... 11(F)
Polymerization, defined 11
Polymethylpentene
  ultrasonic welding
    characteristics .... 57(T)
Polyolefin
  corona discharge treatment 199,
                                                 204(F)
  gas flame treatment .... 213(F)
  induction bonding 100
  modified, corona discharge.............202
  pad transfer printing 154
Polyphenylene oxide (PPO)
  internal coloring for 181
Polyphenylene sulfide (PPS)
  ultrasonic welding ....57(T), 63
Polypropylene ....9(F)
  adhesion to 31
  compatibility with other
    thermoplastics .... 58(F)
  corona discharge treatment 199,
                                    204, 205(F), 206(F)
  cumulative domestic growth .... 16(F)
  feedstock materials .... 11(F)
  hot gas welding 110
  internal coloring for 181
  pad transfer printing 154
  plasma surface treatment ..217, 221(T)

  semicrystalline structure 16
  surface characteristics 25
  surface energy 200
  surface free energy.................... 208(F)
  ultrasonic welding .... 57(T), 63
  vacuum metallization 163
Polystyrene (PS)
  amorphous structure 16
  characteristics 15
  colorant use in 20
  compatibility with other
    thermoplastics .... 58(F)
  corona discharge 203
  cumulative domestic growth .... 16(F)
  feedstock materials .... 11(F)
  G.P., ultrasonic welding
    characteristics .... 57(T)
  impact modified, ultrasonic
    welding characteristics .... 57(T)
  in ABS copolymer 17
  in-house coloring 182
  internal coloring for 181
  polymerization.............................. 11–12
  surface characteristics 25
  surface energy 200
  surface free energy.................... 208(F)
  ultrasonic welding............................ 63
Polysulfone (PSU)
  ultrasonic welding...................... 55, 63
Polytetraduoroethylene
  surface free energy.................... 208(F)
Polyurethane adhesives
  corona discharge 202
Polyurethanes (PUR) ....40, 184(F)
  applications 184
  as conformal coating...................... 189
  cumulative domestic growth .... 14(F)
  soft paint systems........................... 186
Polyvinyl acetate (PVAC)................... 37
Polyvinyl alcohol (PVA) 37
Polyvinyl chloride (PVC)
  amorphous structure 16
  compatibility with other
    thermoplastics .... 58(F)
  cumulative domestic growth .... 16(F)
  dielectric sealing 78–79
  as fixture material for pad transfer
    printing 155–157

as fixturing component material for
   vertical stamping machines..........123
flexible, ultrasonic welding
   characteristics............................. 57(T)
plasticizers..22
rigid, ultrasonic welding
   characteristics............................. 57(T)
surface characteristics 25
thermal stabilizers...............................22
Pot life
   defined 37
   two-component inks 154
Powder coats 186
Power output, of booster horn 76
PPO. *See* Polyphenylene oxide.
PPS. *See* Polyphenylene sulfide.
Precolored plastic...............................182
Pretriggering
   of ultrasonic vibrations 59
Pricing
   OPEC oil price effect on
      plastic prices .... 11, 12(F)
Primary process standard cost............2
Primer, defined...30
Printed circuit boards (PCBs)
   conformal coatings .... 188–189(F)
   shielding, as application of heat
      staking 100
Printing ....193–197(F)
   gas flame treatment theory 209
Propylene ....9(F)
PS. *See* Polystyrene.
PSA. *See* Pressure sensitive adhesives.
PSU. *See* Polysulfone.
PUR. *See* Polyurethane.
PVA. *See* Polyvinyl alcohol.
PVAC. *See* Polyvinyl acetate.
PVC. *See* Polyvinyl chloride.
Pumps, for vacuum metallization......166
PZT. *See* Lead zirconate titanate.

R

Radio frequency (RF) field 217
Radio frequency interference (RFI)
   reduction 161

Radio frequency (RF) sealing 78–79(F)
Recycling
   decorated and/or assembled parts 3, 6
   dielectric sealed plastics....................79
   heat staked plastic assemblies 104
   injection molded parts 2
   vibration welded plastics 51
Regrind............................... 18, 24–25(F)
   defined..24
   effect on plastic part strength ..... 24(F)
   and ultrasonic weldability.................56
Reinforced plastic, colorant use in.....20
Reinforcements 18–19
   advantages and disadvantages 19
Reinforcing agents
   and ultrasonic welding 56
Release agents 165
Remolding..2
Resin grades
   and ultrasonic weldability.................56
Resins
   Amorphous
      and ultrasonics.................................55
      ultrasonic welding..........................63
   crystalline
      and mold release agents.................56
      and ultrasonics.................................55
   hygroscopic 55
   ultrasonic welding.............................63
Resistivity
   material, induction inserting 85, 86(T),
                                                87(T)
Rigidity, reinforcement effect....... 18–19
Rotary index tables..................122, 147
Runners................................227–229(F)

S

Sacks ....12, 13(F)
Safety standards 6
Salt spray tests
   vacuum metallized parts................. 169
SAN. *See* Styrene acrylonitrile.
Saran, surface free energy............ 208(F)
Scissioning 215
Scotch-Weld 2219 epoxy 220

Scotch-Weld 3549 urethane 220
Scrap 1
Scrap cost 3
Screen printing .... 196(F)
  gas flame treatment level
    requirements 207
Servomotors 193
S-glass 19
Shapes .... 11–12, 13(F)
Shear, bond strength .... 41–42(F)
Shelf life, defined 37
Silicone-based materials 40
Silicone rubber
  die failure causes .... 127(T)
  as die material for hot
    stamping 124(F),125–128(F,T), 134
  dies compared to metal dies ..... 130(T)
  heat loss .... 128(T)
  notes on specifications .... 126(T)
  product comparison .... 125(T)
  product specifications .... 126(T)
  properties of formulations .... 128(T)
  as roller material for roll-on
    decorating of foils .... 136, 138(F)
Silicone rubber transfer pads ... 143(F),
                145(T), 147(F), 148–149,
                            151–153(F)
  base materials 153
  hardnesses 151–152
  shapes and applications ..... 151, 153(F)
Silicones 37, 55
  as conformal coating 189
Silverware
  as application of induction
    inserting .... 93–94, 95(F)
Skin depth .... 86(T)
  equation 86
Snap-fit part .... 1, 3(F)
Soaps 234
Soft paint systems .... 186–187(F)
Solvent bonding .... 31, 32(F)
Sound speed in horn material (C) 76
Soybean by-products .... 11(F)
Spark breakdown 202
Specific gravity
  silicone rubber formulations .... 128(T)
Specific heat

induction inserting, metal to plastic
    bonding .... 85, 86(T), 87–91(F,T)
Spectra, plasma surface treatment 222
Spin welding
  advantages and disadvantages 47
  applications 47
  counter-rotation .... 45, 46(F)
  description ...... 45–47(F), 48(F), 49(F)
  equipment .... 45(F)
  flow rate compatibility 47
  joint area design .... 47, 48(F)
  materials criteria 46
Spray and wipe painting .... 194(F), 195
Spray-on protectorates 22
Spray painting .... 189, 190, 191(F)
Stabilizers 18, 21–22
  thermal 21–22
  ultraviolet light 22
Stakes
  flush .... 61, 62(F)
  hollow .... 61(F)
  knurled .... 61(F)
  spherical .... 60(F)
  standard flared .... 60(F)
Stall point (S_1) 75
Stationary inserter variable
  x-y table .... 83(F)
Stationary table
  variable position inserter with
    robotic arm .... 83, 84(F)
Steels
  as die material for hot stamping ... 124,
                        125(F), 130(T), 134
  as fixture material for pad transfer
    printing 155–157
  as fixturing component material for
    vertical stamping machines 123
  induction inserting, metal to plastic
    bonding .... 86(T), 87(T),
                    88(F), 90(F), 91–92
  threaded inserts for induction
    inserting .... 79–80, 81–84(F,T)
Strain .... 15(F)
  defined 15
Stress .... 15(F)
  defined 15
Styrene
  electrolytic plating 171

vacuum metallization 163
Styrene acrylonitrile (SAN)
   compatibility with other
      thermoplastics .... 58(F)
   ultrasonic welding .... 57(T), 63
Styrene-maleic-anhydride
   compatibility with other
      thermoplastics .... 58(F)
   ultrasonic welding
      characteristics .... 57(T)
Styrenics
   cumulative domestic
      growth .... 16(F)
Substrate
   adhesive selection factors 34
   bond strength 41
   defined 30
Sulfones
   compatibility with other
      thermoplastics .... 58(F)
   ultrasonic welding
      characteristics .... 57(T)
Super glue 40
   bond strength 41
Supersil
   description and
      application .... 125(T)
   as die material for hot stamping
      vs. metals .... 130(T)
   mechanical properties .... 128(T)
   physical properties .... 128(T)
**Surface characteristics of
   plastics** .... 25–27(F)
Surface energy .... 199–200, 201(F)
   gas flame treatment
      necessary .... 207, 208(F)
Surface free energies
   of solids .... 208(F)
Surface preparation .... 199–224(F,T)
Surface tension .... 26–27(F),
                        199, 200
   gas flame treatment
      necessary .... 207, 208(F)
   of liquids .... 208(F)
Synergism 171
   defined 17
Syntactic foam 24

T

**TAFA arc spray moldmaking
   technique** .... 195–196(F)
TAFA gun .... 195(F), 196
Talc, effect on ultrasonic weldability ..56
Teflon
   as fixturing component material for
      vertical stamping machines 123
   as fixture material for pad transfer
      printing 155–157
Tefzel 200
   lap shear adhesive strength after
      plasma surface treatment 220
Tensile, bond strength .... 41–42(F)
Tensile properties .... 15(F)
   reinforcement effect 18–19
   silicone rubber formulations ..... 128(T)
Terpolymers 17, 171
   ultrasonic welding 64
Tetra-Etch
   lap shear adhesive strength after
      plasma surface treatment 220
Tetrafluoromethane 218
   added to oxygen plasma 217
Thermal conductivity
   induction inserting, metal to plastic
      bonding .... 85, 86(T), 87–91(F,T)
Thermal transfer marking 6
Thread fastening systems 40
Thermoforming 1
Thermoplastics .... 12, 13(F)
   adhesives .... 37, 39(F)
   amorphous .... 13(F), 14, 15–16, 25
      ultrasonic welding .... 57(T), 63
   crystalline .... 16(F)
      ultrasonic welding .... 57(T), 63
   defined 14
   ductile .... 15(F)
   elastomer 14, 15
   fillers 20
   flexible 15
   forms .... 14–15(F)
   hot stamping 115
   induction inserting .... 79–98(F,T)
   pad printing 141
   plasticity 15

Thermoplastics (continued)
    powder coats 186
    rigid .... 15(F)
    semicrystalline .... 13(F), 14, 16(F)
    spin welding ....45–47(F), 48(F), 49(F)
    ultrasonic insertion .... 69–73(F,T)
    ultrasonic welding ..... 54, 57(T), 58(F)
Thermoset plastics .... 12–14(F)
    adhesives .... 37, 38(F), 39(F)
    colorant use in 20
    cross linked .... 13(F)
    cumulative domestic growth .... 14(F)
    fillers 20
    flash removal not necessary 229
    hot stamping 115
    as multicomponent paint systems 184
    pad transfer printing 141, 154
    powder coats 186
Thermosetting process .... 13–14(F)
Thermosil
    description and application .... 125(T)
    as die material for hot stamping
        vs. metals .... 130(T)
    mechanical properties .... 128(T)
    physical properties .... 128(T)
Tipping .... 123–124(F), 125
Titanium
    as ultrasonic assembly horn
        material 77, 78
Titanium dioxide
    effect on flexibility of plastic 20–21
    as white colorant 20
Toluene, surface tension .... 208(F)
Tumbling .... 231–232(F)
2-cyanoacrylate 40
    bond strength 41

U

Ultem 1000
    plasma surface treatment .... 221(T)
Ultrahigh molecular weight
    polyethylene (UHMWPE)
    ultrasonic welding
        characteristics .... 57(T)

Ultrasil
    description and application .... 125(T)
    as die material for hot stamping
        vs. metals .... 130(T)
    mechanical properties .... 128(T)
    physical properties .... 128(T)
Ultrasonic assembly system ...73–78(F)
    booster horn .... 73, 74–76(F)
    converter .... 73–74(F)
    horn .... 73, 74(F), 76–78
Ultrasonic degreasing .... 234–235(F)
Ultrasonic forming 54
    thermoplastics 54
Ultrasonic heading .... 58–62(F)
Ultrasonic insertion .... 54, 69–73(F,T)
    troubleshooting guide .... 72–73(T)
Ultrasonic riveting .... 58–62(F)
Ultrasonic spot welding
    thermoplastics 54
Ultrasonic staking .... 54, 58–62(F)
    selection criteria 59
    thermoplastics 54
Ultrasonic vibration
    applied to runners 228–229
Ultrasonic welding .... 1, 52–73(F,T)
    basic principle 57
    characteristics .... 55–57(T), 58(F)
    equipment .... 4(F), 52–54(F)
    joint design .... 62–69(F,T)
    resins 55
    spot welding .... 61–62(F)
    staking/welding tips 62
Ultraviolet (UV) curable coatings .. 167
Ultraviolet (UV) light
    stabilizers 22, 167
UHMWPE. *See* Ultrahigh molecular
    weight polyethylene.
United Silicone, Inc. 125
Unsaturation (crosslinking) 217, 218
Urethane
    coating cast to shape
        of part 123
    as coating for fixture of pad
        printing 157
    in top coat systems for vacuum
        metallization 167
    not metallized 163
UV. *See* Ultraviolet.

V

Vacuum metallization
  adhesive tape tests 169
  applications .... 161–163(F)
  costs 170
  cyclic temperature test 169
  decorative finishes 163
  description of process 163
  developments 162
  equipment systems .... 164(F)
  humidity tests 169
  overlay coatings 167
  pencil hardness tests 169
  process steps 165–168
  salt spray tests 169
  tests 168–170
  top coatings 167
  ultraviolet exposure tests 169
  vs. electroplating 170
  water immersion tests 169
  zinc die castings 167–168
Valox 310
  plasma surface treatment .... 221(T)
"Value-added" processes 2
Vapor deposition polymerization
  to apply conformal coatings 189
Vectra A625
  plasma surface treatment .... 221(T)
Velocity, maximum, or amplitude 77
Venturi mixer 210
**Versamid 140 polyamide
    (General Mills)** .... 221(T), 222
Vibration dampening 33
Vibration welding .... 50–51(F)
  advantages and disadvantages 51
  applications .... 50, 51(F)
  joint area design .... 51(F)
  material compatibility 50
  principle of .... 50, 51(F)
Vibratory bowl feeders 122, 147
Vibratory stacker feeders 122, 147
Vinyl
  cumulative domestic growth .... 16(F)
  surface free energy .... 208(F)
Volatile organic compounds (VOC) 6

W

Water
  in plasma surface treatment 217
  surface tension .... 208(F)
Water-based ink printing
  gas flame treatment level
    requirements 207
**Waterborne or water-based
    paint systems** 185
Weatherometer 169-170
"Wet-out" .... 207, 208(F)
Wettability .... 199–200(F), 203, 218
  and surface energy of plasma
    processes 219
Wetting 25–26
  defined 25
White glue 37, 40
Wood flour, as filler material 20

XYZ

Xylenes, surface tension .... 208(F)

Zero gas regulator 210
Zinc stearate 55